T0226101

The Vortex and The Jet

Reiner Decher

The Vortex and The Jet

A Journey into the Beauty and Mystery
of Flight

 Springer

Reiner Decher
Department of Aeronautics
and Astronautics
University of Washington
Seattle, WA, USA

ISBN 978-981-16-8030-4   ISBN 978-981-16-8028-1   (eBook)
https://doi.org/10.1007/978-981-16-8028-1

This Springer imprint is published by the registered company Springer Nature Singapore Pte Ltd.
The registered company address is: 152 Beach Road, #21-01/04 Gateway East, Singapore 189721, Singapore

To all who look up in wonder
to see an airplane fly
and those who ponder
how and why.

Acknowledgements

This story is made possible by many whose contributions include the motivation for me to write this, for clarification of the understanding of the phenomena, for supplying the images and for the time spent by readers who made the effort of reading early drafts. Their feedback was invaluable.

The motivation was provided by Gary Glenn and a number of fellow docents at the Museum of Flight in Seattle. Gary was central to raising questions about what docents tell visitors about the technology involved in flight. My observations about public understanding further strengthened the need to write what you have in your hands. For the encouragement to proceed with this project, I thank my Thursday morning coffee and muffin sharing colleagues Jim Clyne, Dave Cable and Ken Bottini, all docents at the Museum of Flight.

The clarification of the technology was supplied in large part from my own having taught the subject to aspiring students, and many details were provided by practitioners of the art. These include pilots and colleagues. Among them, I acknowledge Ronald Parker and former student Forrest Sims (… by your students you will be taught, as the song goes!).

Critical reading is necessary for many reasons, and to Reed Davis, Gary Glenn, and especially to Stuart Wier, I extend a heartfelt thanks for many improvement suggestions. A big thank is also due to Molly Blank for her generous help with the generation of some of the figures involving vector graphics. Organizations and individuals who contributed the photographs so necessary for a clearer understanding are cited in the image attributions.

Lastly, thanks to my life partner Mary for tolerating my time on this keyboard. It was a good way to spend the time made available by the coronavirus pandemic of 2020–2021.

A Note on Images

As an educator, I rely on images to help illustrate the physics of the subject in this book. The Internet is a wonderful resource for finding illustrations that an ordinary writer might not be able to generate. The sources I used are identified in the image captions, where available. Permission to use them may or may not be available. I contacted and received permission from the owners of many of the images used here. The list of individuals who graciously permitted image use includes Bob Garrard, Ron Parker, Werner Horvath, Ron Kellenaers, Ron Lindlauf and Barry Latter—to them, a grateful thanks. I especially want to thank corporate entities who made images available: The Boeing Company and General Electric. Contributions were also derived from sources associated with US governmental agencies. These include the National Air and Space Administration (NASA and its predecessor, the National Advisory Committee on Aeronautics, NACA), branches of the US military services and civilian agencies.

I sincerely hope that the use of images from the Internet identified as being in the public domain is appropriate because finding original contact information for the photographer was, in a few cases, impossible. The names are identified in the captions when they could be found. Images from the publication *Flight Literacy* were included in this writing, with permission.

To all individuals and people associated with agencies, a sincere thanks for my ability to use images to strengthen arguments made with the written word.

Contents

Abbreviations

ac	Aerodynamic Center
APU	Auxiliary Power Unit
BPR	Bypass Ratio
CET	Compressor Exit Temperature
CPR	Compressor Pressure Ratio
EPR	Engine Pressure Ratio
GE	General Electric
GPS	Global Positioning System
ICE	Internal Combustion Engine
IGV	Inlet Guide Vane
I_{sp}	Specific Impulse
NACA	National Advisory Committee for Aeronautics
NASA	National Aeronautics and Space Administration
NIMH	Nederlands Instituut voor Militaire Historie
NOAA	National Oceanic and Atmospheric Administration
NPR	Nozzle Pressure Ratio
P&W	Pratt and Whitney
psia	Pounds Force Per Square Inch (absolute)
RPM	Rotations Per Minute
SSME	Space Shuttle Main Engine
TET	Turbine Exit Temperature
TIT	Turbine Inlet Temperature
VG	Vortex Generator

Chapter 1
The Vortex—A Journey into Beauty and Mystery

Air museums around the world are displays of the manifestations of the many ideas that went into making it possible for a person to take to the air. The earliest desires to do so are recorded in mythology, the human imagination, and envy of what birds can do. Icarus and Leonardo da Vinci dreamed of flying but, alas, their efforts were not successful. Balloons lifted people into the air using the buoyancy of warm air in the eighteenth century. Just going up was not enough. We want to go somewhere! Serious thoughts about flying using air-moving (aerodynamic) forces also date back a couple of centuries with great minds addressing the possibility.

The intent of this writing is to tell the interesting story of the science of flight and, in part, to dispel some misunderstandings about the topic. The reader is assumed to be interested in aviation generally without formal training in the subject. The story centers on two main topics: the generation of lift by a wing and the generation of thrust by a modern jet engine. Both are the result of the successful development of a wing for controlled flight and the internal combustion engine. While many labored to be first to fly, the Wright brothers succeeded as history witnessed. The contributions made by others should not be underestimated. The history of how we got to our time when non-stop flight half-way around the world is possible is told by many others. One such story told by the author may be of interest.[1]

The curious observer of airplanes would surely enjoy having the mysteries of flight explained in simple terms that do not rely on mathematics so the beauty inherent in successfully doing so can be better appreciated. Placards or descriptive scale models in a museum are limited to explain the function of the artifact in view. Words by personnel available may also not be adequate in clarifying the matter. To answer questions like the above and more detailed ones like: Why is this airplane built like this? or What does this feature accomplish? requires a little study and this narrative is meant to shed some light on such questions.

The endeavor of flight is a technical one and there is no way around that fact. There was physics to be exploited to the point where it could be utilized even if

---

[1] See "Powering the World's Airliners" in the bibliography.

© The Author(s) 2022
R. Decher, *The Vortex and The Jet*,
https://doi.org/10.1007/978-981-16-8028-1_1

detailed understanding was not yet in hand. Structures had to be invented to allow for light-weight "wings" to be built. That necessarily involved understanding the limitations of materials with which to build a wing and a body of something that would fly. Ideas to do this came from observation of nature and a keen pursuit of "let's try this!" It finally came together with the efforts of the Wright brothers who, above all, realized and addressed the importance of control so that flight was possible and relatively safe.

A lesson learned by all who attempted to fly is that the human body cannot exert sufficient power to fly to the degree that a bird can. A solution required an extension of the understanding ushered in by the Industrial Revolution. The steam engine developed then as a means to produce more power than a horse had to be improved upon. It was too heavy. A flight capable engine had to be a small and light-weight package. That step was realized with the invention of the gasoline, and later, the gas turbine engine.

All the basic challenges to flight were met successfully in the first half of the twentieth century. The necessary engine made flight practical. Materials, structures, controllability, and aerodynamics were mastered to the point where the airplanes we use today to travel to all corners of the world are a reality. Aviation was eagerly embraced by the military as well as individuals who can use aircraft of various kinds for commerce, business, pleasure, or sport. The typical air museum will have on exhibit many such craft to amaze us with the ingenuity that was built into them.

To people who are not versed in the sciences of flight, there are many questions that may want answers or, at least, plausibility of understanding. Two such questions stand out and these are addressed by many attempts at explanation: How, or better why, does a wing generate lift? and How does the jet engine work to provide the thrust?

The ability to design, build, and operate an airplane safely is a firmly rooted in technical understanding of the physics of the motion of our invisible air. The functionality of an airplane is not based on magic. In connection with the generation of lift, we will conclude that the vortex is a central feature of fluid motion and is necessary for a wing to function. Another conclusion is that fluid friction, while it may be detrimental to performance, is also necessary. With friction always present, sufficient power has to be provided to overcome the drag associated with flight. An important aspect of our story is that it is challenging to force fluid to flow "nicely" into a region of higher pressure. This is what we ask air flowing over a wing to do after it has provided a region of low pressure on the upper surface of the wing for lift. That aspect imposes a serious limit to what airplanes can do for it involves drag and even the possibility of stall, i.e., a failure to provide lift.

Not only does friction enable lift, it also plays a role in drag associated with air rubbing along an airplane's surfaces. It also plays a critical role in determining the pressures we might have thought we designed into a wing. Both friction and pressure forces on an airplane surface are important components of resistance to motion that has to be overcome with an engine. Hence it is important to understand their nature. We extend that exploration to look at the heating of airplane surfaces experienced by flight at very high speeds. Spoiler alert: it is not friction that causes the body of the Concorde or the SR-71 to be hot when landing.

Understanding the geometric aspects of flow is one thing. Turning that to a knowledge of forces experienced by a wing requires the ability to relate velocities to pressure. We consequently delve into the intricacies of the relationship of fluid motion on the pressure that may be experienced by a surface that tries to coerce the air to follow a prescribed path. The Bernoulli principle,[2] a common way expressing changes in velocity to changes in pressure, has limitations that are "easily" overcome with a better understanding of fluid motion physics.

Use of the word "fluid" here is to alert the reader that we deal primarily with air as a medium because flight takes place in our atmosphere. Water is also a fluid medium. It has the advantage of providing us with many observable situations, a luxury that air provides less readily because it is invisible and transparent. Thus, thinking about water can be very helpful because these two media share a number of similarities in the way they behave.

This discussion will largely be limited to subsonic aerodynamics, the kind we experience stepping in a small sport airplane or a modern airliner. In subsonic flight shock waves will be involved in regions of locally supersonic flow. These waves play a role in placing limits on the performance of wings.

In the second part of this writing, we will extend our understanding of wing performance to gain insight on propulsion and, specifically, how a jet engine works. There are lots of little wings in that engine and they work well to deliver amazing performance in many of the airplanes we fly. Our understanding of lift on the wing is applicable to what happens or can happen in a jet engine.

To be honest and unapologetic, I will use words related to the physical phenomena that have specific and precise meanings. These can be investigated further in the technical literature. To the best of my ability, I will minimize the use of mathematical statements called equations. They can intimidate the lay reader, but for the engineer, they are visually and intellectually powerful ways to express relationships. In order to allow for further investigations by the reader, I will describe simply and accurately the various concepts necessary for the discussion. Hopefully these words will help.

To describe properties of air and other fluids, I will occasionally use symbols. Such use will be limited in the text, but more complete in an appendix. Symbols are very convenient shorthand. The subject of this story covers aspects of two separate disciplines that necessarily merge for our look at flight: aerodynamics and thermodynamics. Each carries its own language and symbols. On several occasions, a specific symbol may be used to describe different quantities to be introduced in the proper place of our story. Examples are: (1) Velocity V and its components in various directions u, v, and w share their symbol use with u also used to describe internal energy and v also used for specific volume, (2) The symbol C_p is applied to a non-dimensional pressure coefficient and the air specific heat at constant pressure, (3) q may be used to denote dynamic pressure or heat transferred per unit mass, and (4) w is used for downward velocity (see (1) above) and work in the thermodynamic

---

[2] After the Swiss mathematician and physicist Daniel Bernoulli (1700–1782).

sense. These possible sources of confusion should not be a problem for the reader because the context is usually clear. In practice, these quantities are seldom (here, never) used in the same context.

Finally, the need for quantitative description of the physical world, has led to the use of two systems of measurement, the so-called English System and the S-I (metric) system. Historically, the English system has been and still is in use in the United States and in the aviation world even as the world uses a much more practical and much more widely accepted metric system for most other purposes. I, having grown up with the English system, will describe quantities in that fashion and hereby extend sincere apologies to readers whose familiarity with the metric system might force them to think a little longer about the quantities described.

Chapter 2
The Vortex and Wing Lift

Observing the birds always made it obvious what was needed: a surface and movement of the air. Through the ages, experience with kites and similar man-made devices substantiated the basic understanding and, in more recent times, motivated attempts by men to fly.

In the early days of aviation, late in the nineteenth and early twentieth centuries, knowledge of what a wing can do to obtain lift was limited. The physics we understand today was not available. Early on, the lift per square foot of wing area that could be achieved later was limited by the design of *wing airfoil sections*, the shape of the hardware designed to deflect air nicely. The first generation of aerodynamic surfaces were thin and delicate because they had to be light in weight, made with flexible fabric, and not able to produce a lot of lift. In order to get a sufficient total lift from the wing, the wing had to be large in area. This presented another challenge in that large span wings were hard, if not impossible, to build with the materials that were available while staying within weight limits. A brilliant solution was to employ, as the Wright brothers did, a box-beam approach to building the wings where the top and bottom surfaces form two lifting surfaces of the wings and the other two (front and rear) sides were reduced to vertical struts and wires to carry the shear and bending loads and thus let the air flow around the lifting surfaces. This was the classical design of a biplane (Fig. 2.1). Knowledge of airfoil and wing design and construction technology improved and with the desire for higher speeds, the biplane eventually became a historical artifact.

A look at the current state of the art in wing design reveals that a number of features on the biplane will be replaced, if not eliminated, by improvements in technology: the struts and wires, one of the wings, the rather sharp leading edge, and the materials with which the wing is built. While the internal combustion engine that powered early airplanes will survive, it was supplanted for large and fast airplanes by the jet engine.

How a modern wing provides lift is well understood by professional engineers and the public should rest comfortably with the thought that magic is not involved.

© The Author(s) 2022
R. Decher, *The Vortex and The Jet*,
https://doi.org/10.1007/978-981-16-8028-1_2

Fig. 2.1 Top: 1902 Wright Brothers glider undergoing testing (Wikimedia Commons GPN-2002-00125). Bottom: DeHavilland DH.82 Tiger Moth biplane introduced in 1932 (Photo: NIMH)

Flight is a risky undertaking only when pilots and ground personnel, who are, after all, human, make decisions that have the very small potential of being bad.

A first question is central to flight: How does the wing function? There are curious stories about what causes pressure differences to be experienced by the top and bottom surfaces of the wing and the speeds of air over these surfaces. Some arguments are thought to require that two molecules of air separated at the leading edge have to be reunited at the trailing edge. This is nonsense even if they include vestiges of reality. The pressure differences exerted by the moving air on the upper and lower surfaces of the wing are real and the question may be better stated as: How do these pressure differences arise?

Engineers and scientists work with models. These models may be in their minds, they maybe physical (like a wind tunnel model) or, most commonly, they are mathematical. The mathematical model is very useful because it is quick to be used and can be changed to incorporate a new or better description of the physical phenomena involved. Nowadays such work is almost always done on a computer. We will leave mathematical modelling to the engineers and consider just the physical modelling of air flow, that invisible medium in which flight takes place. We will be unable to avoid description of mathematical relationships entirely. They will have to be invoked and we will do so in hopefully simple easy-to-understand terms.

Our modelling will rely on sketches to represent the basic phenomena because visual illustration can be very helpful to augment a verbal narrative. Or, as somebody once said, a picture is worth a thousand words. Our first question—how does a wing generate lift—is often raised by visitors to flight museums, by students of aviation, and the lay public generally. The answer is very simple if one looks at the big picture but, the details are more complicated.

Let's look at the big picture and take the wing to be some mysterious device that causes air flow onto the wing to be deflected downward. That simple fact means that air that had no velocity (or better *momentum*,[1] our first abstract idea) in the downward direction has been acted upon by the wing to have downward velocity (called downwash). The high-lift system shown in the Fig. 2.2 should make that function clear if this wing succeeds in directing the air as suggested by the hardware. According to Isaac Newton, a force is required to affect a change in momentum and the wing does just that; it creates such a momentum where none was present before the wing came moving along. The reaction force is the lift we seek to understand. That is the easy explanation. Harder to understand are the mechanics of fluid motion and its ability to impart pressures to the surfaces of the wing.

The image in Fig. 2.3 shows the external consequences of an airplane flying in air made visible by clouds. What is apparent is that two rotating masses of air are involved; rotating in opposite senses. Each such flow feature of rotating air mass is called a *vortex*. It has a center and its influence decreases further from that center. In

---

[1] Momentum is the product of mass and velocity. Velocity is a speed with direction.

Fig. 2.2 A wing of a Boeing 727 with all high lift devices deployed (Photo by author, Museum of Flight, Seattle)

Fig. 2.3 An airplane flying through a cloud layer displaying the downwash and the associated and necessary wing tip vortices (Credit: istock.com: image 1125702712/alextov)

isolation, a vortex flow has *streamlines*[2] that are circular. Because such motion trails from the passing wing, i.e., shed by the wing, they are referred to as *trailing* vortices.

The details of vortex flow will become clearer further on but for the time being, we can view the vortex as similar to a tornado or hurricane. The high (circular) velocities

---

[2] In steady flow, that is the norm for the discussion here, a streamline is a fluid particle path and parallel to the local velocity vector.

Fig. 2.4 A flow deflecting airfoil and a 1909 application by Alberto Santos-Dumont flying his Demoiselle no. 20. The image is provided by Stuart Wier. A similar image may be found at the National Air and Space Museum online archive, NASM 1B35740. This image is flipped left to right from the original for illustrative purposes

near the center can be violent and destructive as indeed they are in storm systems, and they also have to be reckoned with in flight situations. For example, a small airplane following a large airplane that created a strong vortex, say in a landing pattern, can experience an unpleasant, if not deadly, encounter because the small airplane may be rotated by the air in which it tries fly.

2.1 The Bound Vortex

In addition to the trailing vortices created by the passing airplane, there is also vortex-like motion around the wing itself as we suggested when we mentioned wings shedding two vortices. Our first task is to show that this is true and specifically that it is intimately associated with the wing's lift. In short, we will show that the bound and trailing vortices form a system. The first step in understanding the pressures on the wing is to understand the velocity field. The relation between pressure and velocity is set aside for now.

Consider a uniform flow around an object whose function is to create a downward flow. We adopt a viewing platform at rest relative to the device and let the flow arrive uniformly from the left, far away. The picture and our perspective will be in two dimensions in the plane of this sheet of paper. A student might suggest that an airfoil[3] shaped as in Fig. 2.4 would do the trick: turn the flow downward in a smooth fashion. In practice, that turns out to be easy to do, provided we do not ask that the turning angle be too severe.

The totality of the oncoming flow will be divided into two parts, one above the airfoil and the other below. The lower flow between the airfoil and any boundary near or far away will be slowed and, to conserve mass flow of the oncoming freestream, the

---

[3] An airfoil is a rigid shaped object designed to deflect air. Initially, we may take this airfoil to be a very thin sheet.

upper one will be accelerated, at least locally. The idea of mass conservation is that any fluid coming in from the left must exit on the far right. This picture implies that the insertion of the lifting airfoil into the uniform flow field results in the addition of rotation onto the uniform field. An observer would also add that whatever rotational phenomena are involved should be vanishingly small far away from the airfoil.

We are going to argue that this rotation is associated with a vortex connected with the presence of the lifting airfoil and what the airfoil demands the oncoming flow to do. This conclusion is not trivial but let us give it a try.

2.2 Circulation

We can describe the rotational aspects of the flow by drawing a closed path around the far field that includes the airfoil and ask: "what is the *circulation* along that path?" The circulation is the total sum of velocity along the closed path.[4] It is a measure of the rotation of the fluid enclosed by the path. There is zero circulation in a uniform flow because there are portions along the path where the velocity and the path are aligned leading to a positive contribution to the circulation and there will be an equal path length where the flow is in the opposition direction to the path contributing negative circulation. These two elements will cancel out and the conclusion is that uniform flow has no circulation.

If one draws a path around the airfoil, the fact that the upper surface flow velocity is faster than that along the lower surface flow implies that the lifting airfoil must have circulation (rotational flow) associated with it (Fig. 2.5).

Whatever circulation or rotational flow is associated with the airfoil is a feature of its geometry and cannot be associated with the way we measure it. In other words, the circulation must be independent of the diameter of any large circular path we draw around the airfoil. That path length is proportional to the radius (or diameter) of our chosen path. This is possible only if the rotational velocity varies *inversely* with distance from the center of rotation and includes the center. Such a flow field is called a vortex. The circulation is the strength of the vortex. Mathematically, the rotational velocity around the vortex at any point is the circulation divided by the radius to this point, or more precisely, divided by the circumference of the circle. Students of vector calculus would say that the velocity field is curl-free, except where vorticity is located.

2.3 Friction to the Rescue

There is a fundamental problem with this model of a vortex: the implied rotational velocity at the center is infinite because the radius is zero! Nature abhors infinite

---

[4] The engineer or mathematician would call this a *line integral* of the velocity.

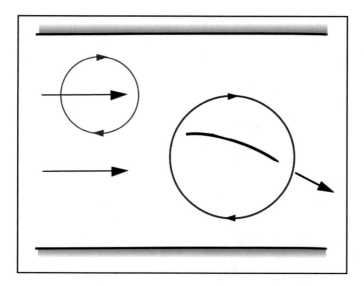

Fig. 2.5 Circulation is the sum of velocities measured along a path such as the circles shown. In a uniform flow (at upper left) circulation is zero. Around the airfoil it is finite. The boundaries shown could be far away or not

quantities so that a mechanism will have to be invoked to prevent that. Real vortices in nature are thereby distinguishable from mathematical ones by the flow behavior near the center. The mechanism invoked by nature is friction. Near the center of the vortex, friction becomes so effective that the fluid there rotates as a solid body. This happens because shearing forces dominate, and the entire central portion of the vortex rotates free of internal relative motion. In such a body the rotational velocity is zero at the center of rotation and it grows larger with increasing radius. Instead of an infinite velocity at the center, we end up with a zero velocity. Notice that we have, so far, not talked about turbulence. In the model vortex with its infinite speeds, it plays no role. In a vortex involving real air, turbulence is the friction mechanism that converts heavily sheared fluid motion to the solid body rotation in the core by means of three-dimensional eddies that decay to ever-smaller size and ultimately to thermal motion at a molecular level, that we perceive as heat.

In a real vortex like a hurricane, the center or eye, of the storm is relatively calm as it passes over a location. At the center, the circular velocity is low and increases as one looks at speeds further from the center. The nature of a real vortex is of two very different characteristics: circular velocity proportional to radius near the center and varying inversely with radius far away from the center. Somewhere between these regions this azimuthal (to use the right term) velocity will exhibit a maximum as the character of the flow blends between the two types of velocity variation described above. In order to describe the strength of a storm vortex, the weatherman would stress the wind speeds in the transition zone because they are highest there.

Fig. 2.6 Satellite view of hurricane Katrina (2005) from space. The image shows the dominant circumferential flow around a core with a modest addition of radial flow and vertical flow between the earth's surface and the upper atmosphere (US National Oceanographic and Atmospheric Administration)

The real vortex is therefore not a singularity (with an infinite velocity) but operates with the rotating part of the flow *distributed over space*. That space could be an area on our page (the rotational core) rather than a point. A way of visualizing this is to think of a vortex as a line, like a wooden broom handle, and when one looks at the details, the handle appears more like a bundle of smaller sticks. These can be spread out over an area and their density as dots on the page may vary. To accommodate that possibility, one may speak of vorticity[5] as a local description of the rotation of the flow. Vorticity over a finite area will look to the observer as a vortex with a singularity (a non-existent infinite velocity point) when viewed from far way. This breakdown of a vortex into distributed vorticity will be very handy to describe the function of the airfoil. Figures 2.6 and 2.7 show two types of vortex structures in nature.

Figure 2.8 shows the schematic variation of the rotational velocity component around a real vortex (lower part of the figure). The upper portion shows the spatial distribution of the vorticity that is collected and relatively uniform near the center and there is none outside the transition zone. In a vortex like a storm system, there are additional velocities, radial and along the rotation axis, that complicate the physics.

The entire flow field, with the exception of the vortex core, is circulation-free and is called *irrotational*. The physical argument is that in order to have local rotation, a torque must have been applied to the material in question. There were no torques

---

[5] Use of the word vorticity is meant to imply that rotating flow is spatially distributed, unlike the word vortex meant to describe the large-scale structure of motion like that of a tornado.

Fig. 2.7 Tornados in the American prairie in various stages of development. The opaque cloudiness is not a direct display of the rotational core flow boundary but rather a measure of the local air speed within the tornado, the temperature, and the humidity of the air (NOAA)

Fig. 2.8 Velocity and vorticity distributions in a vortex in nature. The lower black line shows the azimuthal velocity variation around the axis of the vortex. The upper part of the plot shows the vorticity distribution, uniformly distributed in the solid body rotation part of the real vortex. Going outward, the transition zone is decreasingly rotational and the outer region (past the arrows) is irrotational

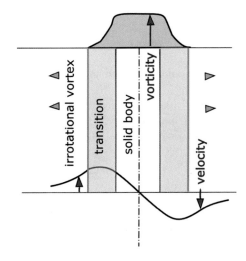

applied to fluid that just came uniformly from the upstream direction. On a real airfoil with flowing fluid, there will be torques applied to parts of the fluid that rub along the surfaces. Friction, as we will see, is the origin of the vorticity involved.

The natural tornado's vorticity is created when two weather systems flow past each other, one above the other. The friction experienced by the atmospheric layer close to the ground slows that air mass and so the stage is set for the air involved to become a tornado. While there are similarities between the natural weather-related vortices and those associated with flight, these similarities do not include the size scale of the vortices, their internal secondary flows and the physics of their creation. The natural vortices are sources of awe and damage. Our concern is focused on the more modest ones associated with flying airplanes.

Fig. 2.9 Superposition of a single vortex on the airfoil and the resulting velocity field. The vortex is centered on the airfoil half-way between points A and B on the heavy line representing the airfoil. The impossibility of a single vortex satisfying the lifting requirement is noted in that at points like A and B, the flow velocity would be *through* the airfoil

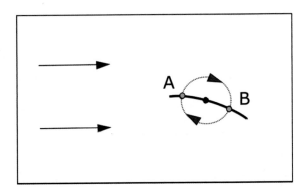

Setting the friction aspects aside for now, we note that a single vortex associated with a lifting airfoil makes no sense. The reason is that a single vortex on the airfoil will have flow velocities involved on some part of the airfoil that are perpendicular to its surface which is, of course, impossible because the airfoil is presumably rigid and does not allow flow through it. Imagine inserting a stationary and rigid, vertical sheet of metal into a water whirlpool and holding on to it. The flow would be severely disrupted, and the vortex would cease to be a single vortex. In all likelihood, two vortices would be created and they would have to move away from the metal sheet. The conclusion must be that a single vortex cannot be located on a rigid surface. Figure 2.9 illustrates this point. Shown there is a single vortex located halfway between points A and B. At the points A and B, the vortex induced velocities would be as noted by the arrows. That is clearly an impossible situation. We need a better model.

So, what to do about modelling the vortex that seems to be necessary to generate lift? We invoke the idea that a vortex can be viewed as a bundle of smaller vortices whose strengths add up to the total. Picture this vortex in three dimensions as a collection of wooden sticks, or better, a bundle of hard spaghetti noodles. If we take this bundle of noodles and spread them out onto the table or some other geometry, the circulation associated with the bundle in its new shape will be the same as if concentrated into round bundle or any other shape. The flow field very, very far from the bundle must be independent of the bundle configuration. Thus, from very far away, the distributed spaghetti vortex is not distinguishable from a tight vortex bundle.

We could postulate that the vortex (or better, the vorticity) on our airfoil is *distributed* along a line co-located with the airfoil in such a manner that the flow resulting from the superposition of the distributed vorticity and the oncoming flow will result in a flow that exactly conforms to the shape of the airfoil. That such a postulate makes sense is suggested by the following thought experiment. Consider that the single vortex in Fig. 2.9 is halved and the two resulting vortices are located at points A and B. The net velocity associated with these two vortices at some location between A and B will be zero because the two *induced* velocities associated with the vortices cancel out. At this point on the airfoil, we have met the requirement that the flow be parallel to the airfoil surface. This thought experiment could be expanded to a very large number of very small vortices of varying strengths, all adding up to the bound vortex strength. This is the distributed strength we seek.

Finding that correct distribution is a formidable mathematical task. Fortunately, theoretical aerodynamicists succeeded in solving this challenge in the 1920s and 30s and since the solution can be found (and agrees with observations), we may conclude that the vortex is indeed distributed in some way along the shape of the airfoil. That allows the ability to investigate airfoil shape analytically. A real step forward.

Chapter 3
Frictionless Air Cannot Provide Lift: A Paradox

Wouldn't it be nice if there was no friction to retard motion we wish to embark on? No, it wouldn't be, for many reasons. We will, however, limit discussion of friction and its role in aerodynamics.

Before we tackle the vorticity distribution along airfoils, consider the flow field that would result if we apply the equations of motion *without friction*. Engineers know how to do this, but you don't have to! It is possible to write the governing equations for flow about an airfoil without the terms necessary to have friction acting between fluid elements. That is a mathematical trick one can use to see what happens. In analyzing flow of honey or motor oil about the airfoil, that approach would be pretty unrealistic. In air, however, it is much better because friction effects are largely confined to the region near boundaries and there, friction does become important enough to require inclusion in any realistic mathematical description ... as we shall see, even as we avoid the mathematics. The 'model' air assumed to be free of viscous friction effects is said to be "inviscid."

The absence of friction means that no torques can be applied to any of the air and hence no vorticity is created. The airfoil will not carry circulation. The mathematical solution with uniform flow far upstream and far downstream being might look something like the streamlines shown in Fig. 3.1. Such a streamline pattern is easy for the engineer to calculate because exclusion of friction effects simplifies the descriptive (differential) equations enormously. In fact, the resulting streamline pattern is identical to one obtained for the constant voltage lines between the top and bottom of the field shown in the figure with the airfoil modeled as an electrical conductor. This similarity is the reason engineers call this *potential* flow.

While nature might seem kinder if the air were to be inviscid (no friction) and the flow field to look this way, the wing designer would not like it. An airfoil in inviscid air would provide no lift because the flow is not deflected downward as Newton's laws require (see Fig. 3.1).

© The Author(s) 2022
R. Decher, *The Vortex and The Jet*,
https://doi.org/10.1007/978-981-16-8028-1_3

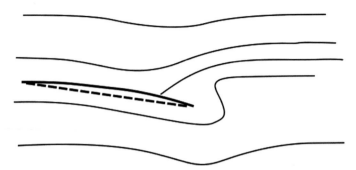

Fig. 3.1 Flow about a flow deflecting airfoil in the absence of flow rotation or vortices. Shown are 5 streamlines including one that impacts the airfoil at the leading edge (not shown) and departs from the top rear surface of the airfoil. The dotted line is the chord of the airfoil

3.1   The Trailing Edge

In a strange turn of events, friction comes into play to rescue the situation: The flow around the trailing edge is not possible because, in reality, the air that has slid along the two sides (particularly the lower surface) of the airfoil has been slowed by friction and thus robbed of the energy necessary to turn the corner at the trailing edge. If the air could travel to the (non-existent) stagnation point on the upper surface it would be going from a region of low static pressure at the trailing edge into a region of the higher pressure at the stagnation point. That is very difficult as we shall see. Instead, the air simply gives up by going in the direction given by the trailing edge angle. Thus, one could say that friction causes the flow to follow the angle dictated by the trailing edge. Two scientists came up with a rule as follows: The trailing edge assures that the wing circulation arises in just the amount required to have the flow leaving the trailing edge in its direction. The German mathematician Martin Kutta and the Russian scientist Nikolai Joukowski came up independently with that rule through observation of the flow physics in the early 20th Century. Another step of progress.

In our aerodynamic world, the friction experienced by the flow on its lower surface is largely responsible for the establishment of the vorticity that provides lift. We might wish we could live without the consequences of friction but, the reality is that we could not. Just picture yourself on a smooth ice sheet trying to go for a walk!

Figure 3.1 is an opportunity to introduce a little more nomenclature. The curved shape of the airfoil *mean line* is normally characterized by the maximum distance (called the *camber*) between the straight *chord* (dashed line) and the mean line, expressed as a few percent of the chord. The airfoil's *angle of attack* is the angle of the chord relative to the freestream direction, that is, the direction of the flow if the airfoil was not present.

Fig. 3.2 Typical vorticity (per unit length) distribution along the chord for an airfoil at an angle of attack. The integrated vorticity per unit chord length is the strength of the equivalent bound vortex. LE and TE refer to the leading and trailing edges of the airfoil

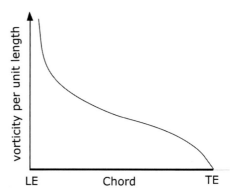

Our understanding is at the point where we (a good mathematician) can determine the vortex strength distribution along any airfoil shape, provided it is reasonable. The calculated vorticity distribution along the chord allows the determination of the velocity distribution everywhere in the field and along the airfoil's two surfaces. At positive angles of attack, velocities on the upper surface will be larger than the incoming flow (the freestream) and those below are lower. The total of the distributed vorticity (i.e., the circulation) is directly related to the lift exerted by the airfoil at the angle of attack chosen (Fig. 3.2).

In summary, the trailing edge installs vorticity on the airfoil and distributes it along the mean line. The distributed vorticity, together with the freestream flow, determine the local flow velocity everywhere in the field, including on both sides of the mean line. A wing section with finite thickness would naturally impose small changes in velocities associated with the thickness, but because the wing is typically thin, the local velocity adjustments for thickness are quite small and we will ignore them here.

How good is this model? For substantiation of what we might measure in a wind tunnel or on an airplane, we must be able to relate the local velocity to the local pressure. Pressure is easier to measure than velocity itself, especially near the surfaces of the airfoil section. The relation between velocity and pressure turns out not to be simple and it will be addressed in due time. Assuming, for the time being, that such capability is in hand, the lift can be calculated for an airfoil at modest angles of attack using our vorticity model and compared to a wind-tunnel measurement. The variation of lift with angle of attack is given very closely as measured in the laboratory, even when airfoil thickness is considered. Another aspect of the model validity is that it correctly provides the location on the airfoil where the force is effectively applied. The engineer calls this the *aerodynamic center* which for most airfoil designs is, in theory and in practice, at the quarter chord location on the airfoil. There, the pitching moment of the wing is independent of angle of attack and can be easily calculated using the methodology used to determine lift. The pitching moment is a torque about a point on the mean line of an airfoil or, in the case of a wing, about a line parallel to the span that must be countered by a torque from the horizontal tail surface.

So far so good, but the model is not perfect. It ceases to be adequate at large angles of attack because the flow fails to follow the prescription imposed by its shape. In ordinary aerodynamic terms, we encounter *stall* of the wing at high angles of attack. The model also says nothing about the resistive force called drag.

The trailing edge function seems pretty straightforward in light of the arguments made above: it must be sharp and point downward. It may be noted that the local vorticity at the trailing edge must be zero, according to the Kutta-Joukowski rule so that the air velocities along the upper and lower sides of the airfoil are identical, in speed and direction.

At this point in the evolution of our understanding of aerodynamics we have the capability to analyze shapes (the airfoil section) we might consider to be the basis for constructing a three-dimensional wing. We are, however, a long way from having the ability to specify the lift that is desired and, with a mathematical model, determine the right airfoil or wing shape. We have not yet even considered that resistive forces, otherwise known as drag forces, will be in play. The road to a practical and well performing wing was not going to be easy! More breakthroughs in the necessary understanding will be made and we will tackle them.

3.2   The Leading Edge

As long as we are playing with the design of an airfoil to generate lift, we might say a word or two about its leading edge. The leading edge plays a different role from that of the trailing edge. First, we recognize that near it, skin friction has little role to play on flow behavior because this edge encounters clean freestream flow. In an airplane application, the angle of attack and the associated lift must be allowed to vary with the needs of a maneuvering airplane. Consider, for example the simple flat plate airfoil at a modest angle of attack. Figure 3.3 shows the oncoming stagnation streamline attaching to the airfoil at the lower surface. When the angle of attack is reduced, this attachment point wanders forward and closer to the real leading edge.

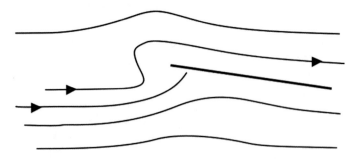

Fig. 3.3 Sketch of the flow about a lifting airfoil. This figure is not unlike Fig. 3.1 rotated 180 degrees

Fig. 3.4 Photo of wind tunnel test with smoke tracers. Flow is from left to right (Image: H. Babinsky, University of Cambridge)

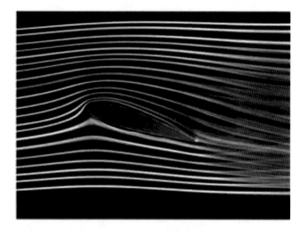

In any case, the flow ahead of the stagnation point must be able to negotiate the flow around the leading edge. A nicely rounded leading edge allows that to take place. Figure 3.4 shows the flow in a wind tunnel with smoke tracing a number of streamlines.

In the later discussion of the jet engine, these ideas about the proper design of wing leading and trailing edges will, amazingly, also apply to the design of inlets and nozzles.

3.3 A Better Model

The representation of an airfoil as a line of vortices in an inviscid (not viscous) flow is a bit simplistic. The mathematicians tackling the problem of developing a decent model of the airfoil were fortunate in that tools were developed to analyze the characteristics of a more realistic airfoil. Among these was the ability to add flow elements to our combination of vortices and freestream to arrive at a good-looking airfoil with finite thickness. For example, a *source* can be added to obtain a flow that looks like the front end of a streamlined body (Fig. 3.5). Another negative source[1] (a sink) can be located behind the source to create a closed, elongated body with flow around it. More realistically, a number of sources followed by sinks behind the point of maximum thickness can be assembled to make a nice-looking airfoil. The beauty of these tools is that they are mathematically relatively simple (for those with that interest[2]) to describe the entire flow field around an airfoil. For simple examples,

---

[1] A source is described by a *radial* flow velocity issuing from a point and varying inversely with radius. See Fig. 7.8 for an example.

[2] A rich resource on the mathematics of the inviscid flow is available on the internet and classical texts under the heading "potential flow.".

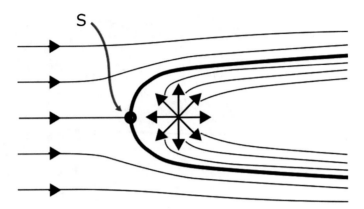

Fig. 3.5 Streamlines resulting from the superposition of a source and a uniform flow from the left. The dividing streamline defines a so-called "Rankine half-body." S points to the flow stagnation point

the results can be generated on a piece of paper, although for complex geometries, a computer will be required.

Chapter 4
Drag, a Nightmare and a Challenge

Lift is, of course, not the only force of interest to the airplane builder because a resistive force must be overcome with propulsion. That force is called *drag* and the subject of much of the design work around an airplane. While lift can be determined from a look at inviscid air flow, drag cannot. That makes for very much more complicated analysis and even a display of the physics.

A classical aerodynamics textbook author will prove the assertion of no drag in potential flow by developing the field equations about a cylinder with circulation. Such a model is built by the addition of a source and a sink, adjacent to each other, and a uniform flow. Such a source-sink combination is called a *doublet* and the flow picture is as shown in the left side of Fig. 4.1. Without circulation, the results show two stagnation points (one where the flow encounters the cylinder at 9 o'clock and another where it departs from the cylinder at 3 o'clock). The addition of a vortex at the center will cause the stagnation points to move from these positions closer to 6 o'clock depending on the strength of the vortex. The fact that the stagnation points relocate near the bottom means the pressure is higher at the bottom than the top, hence there is lift. Because the flow is symmetrical about the vertical axis, there is no drag.

So, what is drag? It has a number of components that are best illustrated by reference to geometries simpler than those of an airfoil.

4.1 The Air Near a Surface

The simplest example to illustrate friction flow kinematics is a flat plate aligned with the flow direction. In a fluid such as air, molecules immediately adjacent to the surface are in mechanical equilibrium with it and must therefore be at rest. This is the no-slip condition we invoked in connection with the Kutta-Joukowski condition for lift. Air molecules further from the surface share dynamic equilibrium with their neighbors: slow ones near the surface and faster moving ones further out. The net

© The Author(s) 2022
R. Decher, *The Vortex and The Jet*,
https://doi.org/10.1007/978-981-16-8028-1_4

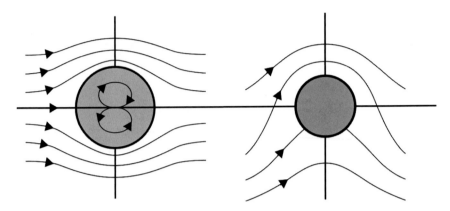

Fig. 4.1 Streamline pattern for potential flow (no friction). Left image is of a source-sink doublet in uniform flow without circulation. The stagnation points are at the 9 and 3 o'clock positions. The image at right is identical with the addition of clockwise circulation by a vortex at the center so that stagnation points move to 5 and 7 o'clock positions

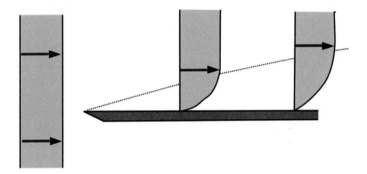

Fig. 4.2 A flat plate with velocity distributions created by initially uniform flow from the left

result is a flow velocity that varies between zero at the surface and full freestream velocity far out. The region of varying velocity is called a *boundary layer*[1] (Fig. 4.2).

4.1.1 The Boundary Layer Comes in Various Thicknesses

The boundary layer is characterized by its thickness. One can choose to describe the boundary layer thickness by the location where the flow is 95% of freestream speed or one can choose 99%. Such a thickness is arbitrarily dependent on the definition. There are more precise ways to describe it. For example, one can speak of the air

---

[1] The concept of a boundary layer is attributed to the famous German aerodynamicist Ludwig Prandtl and dates back to 1904.

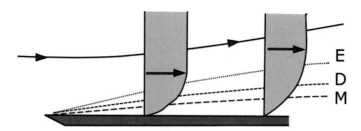

Fig. 4.3 Thicknesses of the boundary layer. Its arbitrary edge is labeled (E). The momentum thickness (M) characterizes the momentum removed by friction forces. The thickness at any point is the drag experienced by the plate to this point. The thicker displacement thickness (D) is a measure of the volume no longer available to the freestream, hence the diversion of the streamline of the flow approaching the plate

volume flow rate that is *not* carried by the region next to the wall and define a boundary layer width representing the volume flow deficit. Such a thickness is called a *displacement thickness*. This region is not carrying the flow that it would carry, absent the boundary layer. As a result, the thin plate is seen by the oncoming flow as a thickening wedge-shaped object. The unavailable volume causes the freestream to be displaced. Thus, the boundary layer alters the effective shape of the surface on which the layer grows. So much for thinking that we prescribe the flow direction by specifying an airfoil shape!

Another example of thickness definition is associated with the momentum that has been removed from the flow by the shearing viscous forces. The boundary layer thickness based on this idea is called a *momentum thickness*. It is numerically smaller than the displacement thickness and is a precise characterization of the boundary layer as far as its resistive force exerted on the moving fluid. Figure 4.3 illustrates these thicknesses and the displacement of the freestream flow away from the plate.

Behind the flat plate, the momentum deficit is a *wake* that can be used to measure the *viscous drag* experienced by the flat plate by examining the velocity distribution (in the flow direction) in that wake. Figure 4.4 illustrates the wake's velocity distribution on and behind the airfoil. The white area behind the velocity distributions is the volume deficit. There is also an associated momentum deficit. Further back, the wake grows in width and the velocities more closely approach the incoming velocity while preserving the same momentum deficit as a measure of the drag the flow experienced.

If the drag force resulting from such a flow situation were the only one, life would be relatively simple for the determination of drag. Real aerodynamic surface are not flat plates nor are they necessarily oriented with the freestream direction. Because of this misalignment of surfaces with the freestream flow direction, both shear *and* pressure forces impact the drag experienced.

The matter is further complicated in that the pressure distribution on an airfoil affects the shear experienced and hence affects the shape and progression of the boundary layer. The pressure history of the boundary layer consequently has an impact on the viscous drag and the effective shape of the hardware. It is not just

Fig. 4.4 The trailing edge of
the plate shown in Fig. 4.3 or
of an airfoil. Flow is from
the left. The points marked
M are the edges of the
momentum thickness of the
boundary layer. To the right
is a sketch of the velocity
distribution in the wake
behind the plate. The
momentum deficit associated
with the broad unshaded area
in the wake is the same as on
the trailing edge

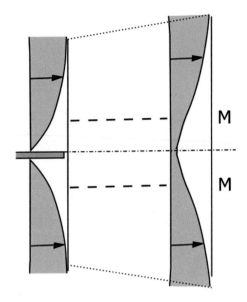

a matter of the freestream flowing along a surface. The shape may even be altered
sufficiently that the freestream flow fails to follow the surface altogether. This is *flow
separation* from the surface.

An extreme example of flow separation is encountered when a flat plate is faced
normal to the flow, as if to stop the flow (Fig. 4.5). There will still be a boundary
layer growing from the impact stagnation point outward along the surface. Here the
boundary layer flowing outward will not have sufficient energy to turn the corner
to re-assemble the downstream flow to resemble what is seen upstream. The flow
simply separates from the edges. In a way, one can think of this as an application
of the Kutta condition. The net result is a high pressure on the upstream side of the
plate and the lower static pressure of the environment downstream of the plate. The

Fig. 4.5 Flow approaching a
plate normal to the flow. The
stagnation point is at A
where the total pressure of
the flow is experienced. A
boundary layer grows from
A to B. The recirculating
region behind the plate is
essentially at flow static
pressure

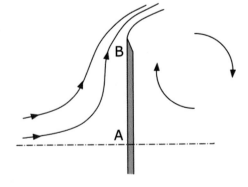

pressure difference leads to *pressure drag*. This is the drag felt by a hand that is held out the moving car window.

4.2 How Important is Fluid Friction?

Air, like any fluid,[2] be it water, oil, or honey, displays a resistance to being sheared. A fluid, by nature, yields to any shearing forces. By contrast, solids do not. The shear stress within a fluid is a force (per unit area, like pressure) on an area subject to antiparallel motion. It is this kind of motion that leads to the rotational nature of fluid affected by friction. The spatial velocity gradient applied to a fluid (picture two flat plates with a fluid between them) and the stress are proportional to one another. The proportionality constant is called the *viscosity*.[3] Such a constant is a valid descriptor of what are called *Newtonian fluids*. Most of the examples cited above are fairly close to Newtonian, especially air. Examples of non-Newtonian fluids are, for example, blood, the honey cited above and many others, for which a descriptive relation between shear and stresses are more complicated because the effective viscosities that are stress-dependent. That is indeed a more complicated world because the relationship between stress and deformation is not simple. It is non-linear. For relatively simple air, life is mercifully easier for modelling its behavior. The magnitudes of the viscosities for the fluid examples above vary substantially and it is quite small for air. How small? you may ask.

4.3 Mr. Reynolds and His Number

For flow issues generally, and especially aerodynamic situations, one would want to have some measure of the relative importance of the dynamic forces being exploited to provide forces on wings and the resistive forces associated with friction. Thus, we might form a (non-dimensional) fraction with dynamic pressure in the numerator and viscous shear in the denominator. Naturally, the viscosity will appear in such a number. With such a formulation for air (and its numerically small viscosity), we

---

[2] In a fluid, molecules move from one region to another carrying with them, dynamic information about mass, momentum, and energy. By this process, properties are transported. Mass transfer is referred to as "diffusion", momentum transfer as "shear" and energy transfer as "heat transfer." These processes can be done on a molecular- or convective levels, which, in aerodynamics, differentiates laminar and turbulent processes.

[3] Viscosity as a numerical parameter has a devilishly complicated set of units. In a discussion of viscous fluid motion, it is often convenient to group the viscosity with the density because they often appear together mathematically, especially in aeronautics. That grouping is called the *kinematic viscosity* and has more civilized units of meters squared per second in the metric system. Near room temperature, the kinematic viscosity has (very) approximate magnitudes of as follows: air (14.6), water (1.0), motor oil (SAE 15W40, 330), honey (2200) all in mm^2/second.

should expect that under normal circumstances where friction effects are small, such a numerical parameter would be very large. In fact, for an inviscid fluid, it would be infinite.

Such a number is called the *Reynolds* number[4] and symbolized by *Re*. An interesting aspect of this number is that it includes a scale length that originated from the measure of the velocity gradient associated with the shear. For an internal flow, such as through a pipe, the diameter is an obvious choice for that length scale. For an external flow like that around an airfoil, the choice for a length scale to describe a viscous flow is a bit of a challenge for a variety of length scales could be used. By far the easiest is to choose some easily identifiable physical dimension of the flow. For an airfoil, the chord is a pretty good choice. Typically, for a whole airplane the wing chord is also commonly used for it is that dimension that best characterizes the flow along the airplane's wing surface. For the localized development of a boundary layer, the path length along which the flow proceeds, a good dimension is the length to any point from the start of the development. In general, the length scale in the Reynolds number is included in the symbol as a subscript, such a "c" for chord. Other length scales may be used. An important one is "x" for path length to characterize the "age" of the boundary layer. This length scale is the relevant one to describe the transition from laminar to turbulent flow to be described further on.

Before going on, it would be good to state the definition of the Reynolds number:

$$Re \text{ or } Re_c = \frac{\rho V c}{\mu} \quad \text{or} \quad [\text{density } x \text{ speed } x \text{ length scale/viscosity}]$$

Here ρ is the air density; V the velocity; c the chord, and μ is the air viscosity. Note the *kinematic viscosity* is $v = \mu/\rho$.

In any system of units, the viscosity of air (μ) is numerically small and, for most aerodynamic purposes, only slightly dependent on the temperature. That smallness will make the Reynolds number in the aviation environment large. Typically, we might expect numbers in the hundreds of thousands up to several million.

4.4 Scale Model Testing

When airplanes did not fly high or fast, similar values of density and speed would apply to a wind tunnel model or to a real airplane. Thus, the model size is the determinant of Reynolds number and the values (characterizing the importance of friction) will not be identical in a wind tunnel model and on a full-sized, geometrically similar article. There is, as a consequence, some risk that wind tunnel data cannot be used to predict full-size airplane performance to a degree that might be desired because the flow conditions are not identical. Under such circumstances, experience

---

[4] After British scientist Osborne Reynolds (1842–1912), a pioneer in the study of viscous flows, among other things.

or mathematical adjustments to model data may or may not be effective in predicting full size airplane performance from data obtained from a wind tunnel test.

For example, a small 4-seat general aviation airplane might have a Reynolds number of four million and a one twentieth scale model would like sport a number closer to 200,000 which lies well under the number of half a million where one might expect the flow to be predominantly *laminar* rather than *turbulent* at the higher number. That wrinkle is taken up in the next section and involves the observation that the boundary layer along a flat plate changes its character when one compares laminar and turbulent flows. We know this because transition to turbulent flow occurs at a Reynolds number based on length of about 500,000. Performance predictions are indeed difficult!

An interesting situation is presented by wind tunnel modeling for a modern airliner. An example might be a 1/20 model of a Boeing 747 run in a transonic wind tunnel. A transonic tunnel operates at higher speed and is specifically designed to operate at Mach number close to one. The effective Reynolds number would be around three million. The number calculated for the real airplane operating in a much lower density and with a larger chord will give a Reynolds number close to 2.5 million. In this case, the compensating scale and density bring the two numbers fairly close. Wind tunnel test data extrapolation is quite modest and carries a correspondingly modest risk to being accurate.

There is some flexibility in choosing wind tunnel parameters that might give better matches between Reynolds number for the wind tunnel test and the real article. The available options are, however, expensive, or technically unrealistic because they violate other similarity rules, specifically those related to compressibility, or flight Mach number. Increasing the size of a wind tunnel (so the model can be larger) may be an option. Another is to operate the tunnel at elevated pressure to increase air density. Finally, gases other than air might be employed. All of these options are costly and not commonly used. The options are limited.

Fortunately, numerical modeling executed on a computer is a great help in allowing airplanes to be designed with reasonably good performance predictions. An observer might even wonder whether the configuration similarities of modern airliners, especially the twin-engine ones, might just have led engineers to identify the best way to build an efficient airplane.

4.5 Turbulent Flow

Back to basics. What about this laminar and turbulent flow issue? It turns out that to make things really interesting, nature added yet another level of complexity to viscous fluid motion. Specifically, the motion of *molecules* sharing dynamic information to result in the velocity profile of a boundary layer works only to a point. Under conditions where it does, the flow is rather well behaved and steady. This is so-called *laminar flow*. Laminar flow is smooth and quiet like a placid river. In many aeronautical circumstances, this lovely state of affairs cannot be maintained as the

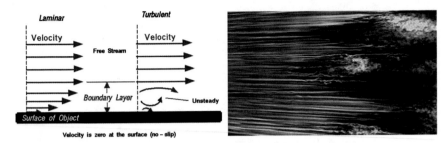

Fig. 4.6 Velocity distributions in laminar and turbulent boundary layer (NASA Glenn Research Center). Figure at right is of a smoke visualization of transition on a flat plate boundary layer viewed from above (From Alfredsson & Matsubara, 2000)

flow proceeds along the solid surface because the thermal motion of air molecules is not up to the task of maintaining the dynamic equilibrium we might wish for. There comes a point where the steady motion becomes chaotic involving unsteady eddies that are large at the edge of the layer and smaller closer to the body. The size variation is imposed by the presence of the boundary. These eddies are very much more effective than the random thermal motion of individual molecules in transferring the low momentum of air molecules near the wall to the freestream air passing by further out. The presence of eddies is associated with a significant increase in skin friction drag. This is *turbulent* flow. The NASA image (Fig. 4.6) shows the nature of the boundary layer for the two regimes wherein it can be found. There are numerous examples of laminar to turbulent flow transitions in nature. The flow from the kitchen water faucet can undergo such a change as can the smoke rising from a burning stationary cigarette.

In a simple flat plate experiment like Fig. 4.2, a Reynolds number based on flow length from the leading edge of about half a million (500,000) will result in turbulent flow. This number may vary a bit depending on the initial turbulence present or the roughness of the surface. Most modern airplanes operate at Reynolds numbers well over half a million (500,000) so that turbulence on aircraft surfaces is normal and unavoidable. The difference in noise levels between the front interior (First Class) of a commercial airliner and that at the rear is a reflection of the more intense turbulence of an "older" boundary layer, i.e., further downstream from its start.

To give some sense of scale for a flat plate experiment in sea level air, the (99%) laminar boundary layer thickness at the point of transition one foot (30 cm) from the leading is approximately 0.1 in. or 3 mm. This experiment would require the air move at about 75 feet/s (25 m/sec), a low speed for an airplane (~80 km/hr or 50 mph in more conventional units of measure). Raising that speed to 750 feet/s (at sea level) would move the transition point toward to a position about one inch from the leading edge and the boundary layer thickness at transition would be very thin ~0.01 in.! One can get a realistic appreciation of the boundary thickness on a real aircraft depicted in this book by examination of the inlet of the Concorde supersonic airliner shown in Fig. 13.6. The varying width space between the wing underside and the inlet upper side is but a few centimeters (about an inch) wide. It varies because the boundary layer has developed further on the inboard side of the inlet structure.

The no-slip condition also applies to turbulent flow albeit in a very thin *laminar sublayer* where interactions prevail on a molecular scale. The thickness of the turbulent boundary is larger than that of the laminar layer. On an airplane, the boundary layer is very thin near leading edges growing to several inches near the aft end of a well-designed body on which it grew. See, for example, the discussion of diverters in connection with jet engine inlets (Chap. 13) and the vortex generators described in the next section.

Description of the behavior of boundary layers on complex body shapes has long been a challenge in mathematical modeling and in wind tunnel testing as the scale dependence of Reynolds number suggests. The descriptive equations are hard to solve because the mechanism of transmitting shear forces changes character between laminar and turbulent flow. Viscosity is no longer the only descriptor of the fluid's frictional properties. The transition also involves a number of factors, among them, how long has the flow experienced viscous stresses, the pressure history, turbulence in the incoming air, surface roughness, etc. The nature of interactions between three-dimensional eddies of vorticity continues to pose challenges for those interested in mathematical modelling.

Having characterized the importance of fluid friction, we can look at more realistic flows to illustrate the nature of pressure drag cited earlier in connection with the general topic of drag. Even though flow around a cylinder or sphere is not of great importance to the performance of a modern airplane, it is interesting and describes the basic idea of the physics around this component of drag.

Under flow conditions described by low Reynolds number (*Re*, based on diameter), the flow around a model cylinder in a wind tunnel may be as shown in Fig. 4.7. As *Re* is increased, the flow becomes unsteady because it is unstable (~100). This is the range wherein the so-called *Karmann vortex street* is formed. It becomes steady again in the region of interest to aviation (*Re* ~ 10,000) and, in this regime, the flow is generally laminar.

The point where the flow separates from the body of the cylinder is important in determining the drag force experienced because it determines the area on which the low (static) pressure (in the wake) acts on the rear side of the cylinder. In this regime,

Fig. 4.7 Steady flow visualization around a cylinder at low Reynolds number (*Re* ~ 26) (Photograph by Sadatoshi Taneda published in "An Album of Fluid Motion, by Milton Van Dyke)

the oncoming flow proceeds around the front side of the cylinder much like flow without friction, although a little viscous drag is experienced there. The pressure is close to stagnation pressure on the front side of the cylinder and is lowest near the 12 o'clock position because of the locally high velocity. The flow separation will occur near there.

At higher Re, (>300,000) the flow becomes turbulent and tends to separate at closer to the 1 o'clock position. The picture of the flow field will resemble Fig. 4.7 with important differences. These include a very turbulent wake reaching far downstream and the absence of any large-scale organized fluid motion.

With a low-pressure wake zone at the rear and a high pressure in the front, the net force resisting the flow motion is the *pressure drag* that dominates the drag performance of objects like cylinders and spheres. The serious student may wish to examine graphically the laboratory result of the reduced wake area by noting that a plot of drag coefficient for flow across a cylinder exhibits a significant drop in turbulent flow (~300,000). This variation in cylinder drag coefficient is attributable to the shifting flow separation point from near 11 o'clock at low Re to 1 o'clock at fully turbulent flow Reynolds number.

On an airfoil or wing, both viscous and pressure drag components are in play. A good aerodynamicist minimizes drag by providing an opportunity for flow to remain attached to the body to recover as much of the higher pressure on the aft end of the body. This is done by increasing the chord of an airfoil-like shape. Such an increase will, however, present a greater surface area which, in turn, increases viscous drag. This is *streamlining* to obtain a configuration with minimum drag. Beyond streamlining is tailoring the surfaces of the airfoil to optimize the pressure distribution for minimal deleterious effects on drag while maximizing lift. Control of the boundary layer evolution in such a way as to remain laminar has proved beneficial. A notable example is the so-called laminar flow wing of the P-51 Mustang during WW II.

The shape of a nice airfoil section, such as in the photo of Fig. 3.4 illustrates streamlining. For some thickness to length (or chord) ratio, there is a minimum in total drag. For example, streamlining reduces the drag from a round wire by about a factor of 10 for an airfoil shape about 10 times longer than wide and with the same thickness as the wire diameter. Biplane builders who necessarily used lots of wires for shear strength in their wings were eager users of streamlined wires when their advantage became clear.

An interesting aspect of the transition from laminar to turbulent flow is that the turbulent layer is more energetic so that it tends to be able to flow further into a region of higher pressure without separating from the body surface. Golf players found that if a ball is dimpled to promote turbulent flow, its lower drag would allow it to travel further!

In wing design, engineers have struggled with means to mitigate the deleterious effect of the boundary layer on a wing. These include energizing the flow near the surface with air from another source (such as an engine) or removing the boundary layer altogether by sucking it away through a porous surface. Such steps are employed, primarily when the need to do so is imperative, because implementation requires power from an engine. An example is on the surfaces of the inlet of an airplane capable of supersonic speeds. There, the interactions between possible shock waves and the boundary layer can be problematic. More commonly, clever use of available air may also be of help. Such methods are incorporated in flap systems used during takeoff and landing, where the flow deflection angles are large. Multiple element flaps are examples of such designs. The use of vortex generators often accomplishes drag reduction on airplane surfaces. Let's examine their functionality.

4.6 Vortex Generators

Since the low energy boundary layer air is the culprit that leads to flow separation from the wing and the associated drag as well as stall, it seems logical that increasing the energy in boundary layer air might allow for better performance on both these counts. Indeed, it does. The tool available is the vortex generator. I'll abbreviate this set of words as VG. That is what the pros call it! It is a very simple vane, usually quite small, at most on the order of inches in size and protrudes into the freestream air past the boundary layer edge. It is mounted on the surface of the area where flow separation is a problem. That may be on the rearmost quarter of a wing's upper surface, near a wing's leading edge for stall control, or on the converging rear section of an airplane fuselage. The shape can vary: a rectangular fin, a triangular fin or a pair of these elements at opposing angles of attack to the airflow.

The function of the VG is to create a trailing vortex from its tip so that exchange between the low energy boundary layer air and the higher energy freestream air is forced. By such means, the deleterious effects of flow separation from the surface are pushed to more severe operating conditions, such as higher angles of attack (Figs. 4.8 and 4.9).

If you thought that the drag picture for an airplane is only about laminar and turbulent flow, or only about pressure and viscous drag, you would be mistaken. The vortices we need for lift also have something to say about drag. Their contribution to drag will also be tackled in Chap. 6.

For now, we have to step back to understand how pressures on airplane surfaces arise from the motion of air over them.

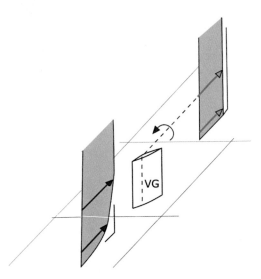

Fig. 4.8 VG pair array on the upper surface on the Boeing 737 prototype at the Museum of Flight. Such VGs may be located closer to the leading edge on a general aviation aircraft because the latter usually don't have leading edge flaps. The sketch illustrates the function of a single VG. Note the velocity deficit approaching the VG (red arrows) is shared with a larger part of the flow behind the VG (green arrows). The no-slip condition applies always

Fig. 4.9 An example of two arrays of VGs on the wing of a Douglas A4D US Navy jet fighter. Museum of Flight display

Chapter 5
Pressure: The Bernoulli Principle and Flow Energy Conservation

To integrate an airfoil or wing into an airplane one must be able to relate the velocities obtained from an examination of the flow's speed and direction, as we have done above, to the pressures acting on the various parts of the airfoil or wing.

The Bernoulli principle is often given as the reason why wings can provide lift and airplanes can fly. It relates pressure and flow speed[1] and thus plays an important role in determining aerodynamic forces. The principle states that in a "nice" flow situation, the pressure rises when the medium is slowed and vice versa. The invocation of this relation that we shall state formally as we proceed, is useful because it describes how changes in velocity relate to local pressure. So, where does this principle come from and is it useful for all situations? We will see that it is valid for incompressible flow, i.e., for the description of low-speed air flow or flows of water generally. To cover its application to aviation, we must keep an open mind because airplanes can easily fly in speed regimes where the air is a compressible medium and, it turns out, the compressibility of air has profound influences on the performance of wings and airplanes.

The discussion to follow is limited to air and water. These fluids are involved in situations where the forces for lift and propulsion are commonly of interest, hence the focus of our attention.

The task of relating pressure to the flow conditions is unfortunately not as simple as it may seem. Our approach will be to describe the basic underlying physical law, a law being an observation that has never been seen to be violated. We will explore it in circumstances where the flow is compressible (as it generally is) and apply it in special situations where it may be simplified as incompressible.

The first task is to address the nature of compressible media. This necessarily involves discussion of quantities that describe the state of air, i.e., its properties,

---

[1] The more general form of the energy balance statement, from which the Bernoulli principle is derived, includes accounting for potential energy changes. That feature must be included in a discussion of larger scale flows such as, for example, those associated with weather. For flows associated with flight, potential energy changes play a vanishingly small role so that accounting for them is unnecessary.

© The Author(s) 2022
R. Decher, *The Vortex and The Jet*,
https://doi.org/10.1007/978-981-16-8028-1_5

specifically pressure, temperature, density, etc. To the possible chagrin of the reader, this also necessarily involves other air properties that are affected by these parameters. Discussion of such properties involve abstract concepts that may strike fear in the reader. The hope is that it does not. The purpose of this writing is to provide light so that the merit of the conceptual ideas may be appreciated. It should be of some comfort that it took mankind until late in the Industrial Revolution to fully grasp these concepts and utilize them. And did we ever!

5.1 Air is Compressible (Sometimes)

What do we mean by "compressible?" Simply stated, compressibility is related to the fact that pressure changes or temperature changes can lead to fluid *density* changes. For air, these three quantities are related to each other in a very simple form: a *state equation* ($pv = RT$) that many students will have encountered in their science studies. The letters stand for pressure, volume per unit mass,[2] and temperature. The constant 'R' is there because we were never smart enough to invent a system of property units to measure *p, T,* and *v* so that $R = 1$. R does carry units that the reader can look up. The temperature and pressure must be stated in values relative to their absolute zeros. Zero temperature is where thermal motion of molecules ceases and zero pressure is called a vacuum. For an incompressible medium like liquid water, a state equation is not particularly interesting because the density (and its inverse, specific volume, *v*) is fixed for all practical purposes.

5.1.1 An Aside: How Good is Air?

A serious student of the matter will notice that we (aerodynamicists) use the ideal gas description of the gas, namely $pv = RT$. There are circumstances where use of this relationship may have to be revisited. That caution also applies to the other constants that will described in connection with other properties, the specific heats of air in particular. The concern arises when the air's thermal energy depends not only on temperature but also on the pressure. Conditions where our simple descriptions might have to be modified are for situations where the air's oxygen and/or nitrogen are near condensation, i.e., very cold. In such states, gases are described as vapors. Near condensation, intermolecular forces play a role and thus result in a pressure dependence of the internal energy. On the high temperature spectrum end of possibilities, such as conditions reached on vehicles re-entering the earth's atmosphere, air molecules break into atoms and even lose their electrons. The radio silence on re-entry is a manifestation of the ionization[3] of the air surrounding the vehicle and the resultant inability

---

[2] Also termed *specific volume* and is the inverse of *density*.

[3] Ionization is the process of breaking down an electrically neutral particle into a positively charged ion and one or more free electrons.

of radio waves to be transmitted through such ionized air. Fortunately for the study of conventional aerodynamics, air, even air with a small amount of water vapor, behaves as a perfect[4] gas and we can safely ignore all these worries about more complex descriptions of air.

The "nice" flow conditions we consider here are flows along a streamline where speeds may vary. A unit mass of air undergoes a process that will be reflected in the changes in the gas properties. The question has to be: What properties describe the process experienced by a chunk of air as it changes speed? Measures of the energy in the flow and the related temperature and pressure are easily seen to vary. The correct measure of thermal energy is a property of the air and will be identified shortly. Is there another property involved? Perhaps conserved? The answer involves delving into questions about *reversibility*. In other words, after an increase in speed and a later equal decrease, are the properties describing the air altered or not? Fortunately, under many circumstances of interest to an aerodynamicist, the air that experiences velocity changes by flowing over an airfoil and is returned to the same pressure it had before the airfoil came along is not (measurably) different in density or temperature. It is said to have undergone a reversible set of processes.

The world is full of *irreversible* processes, but aerodynamic ones of interest to us are fortunately not numerous. They can, however, be challenging. Examples of everyday irreversible processes include the cooling of the cup of coffee on your desk: There can be no realistic hope that the heat that left your coffee will return from the heat in the room to rewarm the coffee. You will have to expend a bit of a valuable resource: the electricity to run the microwave oven. Similarly, all mixing processes are irreversible. Have you ever seen the sugar or cream in your coffee unmix from it without action on your part? Probably not. No sailor stranded at sea has ever been saved by the spontaneous removal of salt from the sea water so he can drink it.

5.2 Entropy (Arrrgh! @*#*!)

A reversible process in air along a streamline conserves a quantity (actually a gas property) called the *entropy* and the process is termed *isentropic* (entropy conserving). In much of aerodynamics, flow changes are indeed isentropic. This is really good because high performance requires that as little of the energy that we invest in traveling through air should be left in the environment as a monument to our having done so. In reality, entropy is not conserved and is instead produced, when friction is involved or when irreversible phenomena such as shock waves are experienced. Both of these aspects of flow are of great interest for an aerodynamicist. For that reason, we will consider them in detail. A process is also not isentropic when heat transfer is involved. While the latter aspect is not in play when looking at streamline

---

[4] A gas is *thermally perfect* when specific heats depend only on temperature. If the specific heat can be assumed constant, the gas is described as *calorically perfect*. Air, under many circumstances, is perfect in both these senses.

flow over an airfoil, it is very important when the flow through a jet engine combustor is examined. Because flows that do *not* involve heat transfer are so common in aeronautics, they are given a special name. They are called *adiabatic* flows. With all that vocabulary in hand, we can summarize our findings with the statement: entropy is conserved when the flow is adiabatic and reversible, otherwise not.

The concept of entropy as a property is a profound one that challenges students in the study of thermodynamics. Understanding this idea, first expressed by the German scientist Rudolph Clausius in the mid nineteenth century, was a scientific tour-de-force whose impact on engineering is hard to overstate. It is usefully applied to the design of engines in particular and answers, in part, questions like: How much of the heat we might supply to an *engine* can be converted to useful work? This was the central question posed by Sadi Carnot, an officer in Napoleon's army. Part of his genius was recognition of the nature of heat and its convertibility to useful work. His inquiry led to significant progress toward the practicality of steam and other engines in the century and a half that followed his life.

The concept of entropy is very powerful indeed, applying to notions of chemical equilibrium, and even speaks to the cosmic possibility of life, among many other situations. Many, many books have been written about entropy and its significance in connection with the so-called Second Law of Thermodynamics since its formulation late in the nineteenth century. Food for thought is that the Second Law is a disarmingly simple and a seemingly obvious statement: heat cannot flow from a cool place to a hotter one! It is a law because it has never been observed to be violated.

We are concerned (for now) with the narrower problem of adiabatic air flow. The mathematics of entropy conservation can be shown to allow a relationship between the descriptive variables of a gas as it undergoes an isentropic process. For our purposes, let us not delve into such details, save to note that such a description exists and is in fact representative of real processes under many but limited circumstances. The mathematical statement relating air pressure and temperature that is imposed by the isentropic process is quite simple. It says that a changing (absolute) pressure will result in a change in (absolute) temperature raised to the ($\gamma/(\gamma - 1)= 3.5$) power. The constant γ (=1.4)[5] is derived from properties of air and is related to the fact that air is primarily composed of the diatomic molecules, oxygen and nitrogen. Its numerical value is highlighted further on. For the interested reader, a mathematical derivation for this relation between pressure and temperature is given in Appendix B.

We return to the question: how do we quantify compressibility? The basis is the understanding that Newton's laws of motion (the various forms and consequences of $F = ma$) can be written in differential form, i.e., for small changes in properties and flow conditions. Such a formulation identifies the parameter that quantifies the importance of compressibility in a reversible process. Without going into algebraic details (professors torture students with this and the details are given in Appendix B), one can show that a fractional change in density is always proportional to a fractional change in local speed. The proportionality constant involves the ratio of local speed to the speed of sound, the *Mach number, M*. The constant is ($-M^2$). To talk about

---

[5] The parameter γ is called the specific heat ratio.

"low speed" flow, we specifically mean low Mach number flow, say values under M ~ 0.3. For example, at this Mach number, a 10% *increase* in local speed results in a ~1% *decrease* in density, which may indeed be negligibly small, but at the flight speeds near the speed of sound, compressibility effects must be considered for an accurate description of the flow field.

The role of the speed of sound in matters relating to compressibility makes sense because it is the speed with which information spreads from one place to another in the flow field. Thus, wave or information propagation is central to the ability of the flow to adapt (or not) to externally imposed condition changes. Air molecules move with thermal speeds, bounce into each other, and share dynamic information. That implies that the speed of sound in air depends only on the temperature. It does not depend on either the local air pressure or density, nor altitude as such. The temperature in our atmosphere does vary with altitude and with it, there is a corresponding variation in the speed of sound.

Consider some examples with a wide set of descriptive circumstances. Both water and air are fluid media and our review of airfoils is equally valid in air and water. The speed of sound in water is very large, about 1450 m/s or over 3200 mph. That sound speed is so high that for practical purposes in realistic water flow situations, like the motion of ships, the ratio of flow speed to pressure wave propagation is very small. Water is nearly incompressible because the Mach number-like parameter describing any realistic situation is very, very small. Our concern is primarily with air in which compressibility effects will be significantly more important. Importantly, we note that the sound speed in water is unrelated to the surface wave speed[6] we describe below in connection with the "hydraulic jump" as an analogy to a shock wave in air. Watercraft dealing with surface waves are involved in a distinct area of study called hydrodynamics where surface waves can be central in importance.

5.3 Energy Conservation on a Streamline

So, what about Bernoulli? We have to get back to basics and discuss another concept that a student usually gets exposed to in a course in thermodynamics. We will do here what is necessary in a hopefully in a comprehensible way.

A basic concept involving variables that describe the state (i.e., the pressure, temperature, etc.) of the air as it undergoes velocity changes is an energy conservation statement, or what is formally termed an energy equation. In a moving fluid there are two forms of energy (per unit mass) when we justifiably neglect changes in potential energy. The first is kinetic energy ($1/2 \, V^2$) and the other is thermal energy; here V is the speed of the air.

The thermal energy is the (randomized) kinetic energy of the atoms or molecules of the air and is measured by the (absolute) air temperature. The absolute temperature

---

[6] The wave speed in water depends on the depth of the water. For example, in a pool of water 1 (10) foot [30 cm (3 m)] deep, the wave speed is about 5.6 (17.9) feet/s or ~2 (3) m/s.

scale is not in common use so that it must be determined from knowledge of absolute zero temperature. On a Celsius or Centigrade scale, water ice (at one atmosphere pressure) freezes at zero degrees C or 273 K (degrees Kelvin) which corresponds to 32 F or 492 R (degrees Rankine) in the English system. The Kelvin and Rankine temperatures are *absolute* scales. The thermal energy associated with molecular motion is formally called the *internal energy* of the fluid and typically described by the algebraic symbol, u.

5.4   Enthalpy is a Property. Really? Really!

The introduction of the word 'energy' in connection with random motion of molecules is an opportunity to digress to explore the nature of heat and energy. Formally, the energy conservation statement is called the First Law of Thermodynamics. Its formulation does two things: it identifies specifically the energy of a substance as a property and defines the nature of heat. The nature of heat puzzled the early investigators to a great degree. The First Law of Thermodynamics allows us (modern engineers and scientists) to state clearly that heat is not a property, but it is that which travels from a hot body to a colder one. This may sound like a mind-game, but it is part of a necessary construct that allows understanding of the nature of heat, work and all that follows from exploitation of that understanding. This is just the topic that Carnot wrestled with. One might wonder just what is a hot or colder body? That, to a modern mind, involves the concept of temperature, that is laid down as the Zeroth Law of Thermodynamics: there is such a thing as temperature! The concept of temperature and its connection with heat was not nearly as well understood more than a century ago as it is today. Consider, for example, the confusion contributed by the reality of phase changes such as melting ice or condensing steam: heat is involved without temperature changes. The zeroth law also points to the existence of the absolute zero in temperature. But enough of the digression into theoretical worlds and let's return to the real one.

The matter of describing the energy equation (or First Law) for a moving fluid is somewhat complicated by the fact that adjacent fluid elements do work on one another as velocity changes are imposed. Picture, for example, the compression of air by a bicycle pump. The air in the pump space is compressed, that is, work is put into the confined air (with a force that reduces the volume) and that results in an increase in temperature. In an antiparallel way, carbon dioxide gas in a fire extinguisher does work to push out what leaves the cylinder when you pull the trigger and that results in a decrease in the temperature of the gas remaining in the cannister. In a flow situation, the work is put into (or removed from) any (and all) elements, not by a piston, but by the pressure forces from its neighbors. Such work interactions affect the state of the gas, specifically its internal energy. An accurate energy conservation statement must account for this work. Work exerted in this way is called *flow work* and is described by pressure times volume (pv, the thermodynamic equivalent of mechanical force times distance). This term is per unit mass because the quantity 'v' is the volume per unit mass.

Making life easier for aerodynamicists, the product "pv" is also linearly related to the temperature through the state description of air cited above: the ideal gas law. The energy conservation statement that applies to a moving fluid is therefore not just about the internal energy (u) but a new quantity consisting of that energy and flow work. This quantity is the *enthalpy*[7] (symbol h, defined as $u + pv$) and, fortunately for air, it is also proportional to just the temperature. It is this measure of thermal energy content in the element that is traded with the flow kinetic energy in our ordinary stream tube. If one defines something called the *total enthalpy* (symbol h_{tot}) of the flow as the sum of static (proportional to temperature) enthalpy and the kinetic energy, then it is the total enthalpy that is conserved in a streamline-like (adiabatic) flow. With a total enthalpy thus defined, the notion of *total temperature*, namely, the temperature that the fluid would have when it is brought to rest, stumbles right out. In algebraic form, the energy conservation statement for our circumstances is

$$h + \tfrac{1}{2} V^2 = h_{tot} \text{ is conserved.}$$

The concept of total temperature is a very useful tool to describe the state of a moving gas along a streamline. The understanding to this point allows that quantity to be written in terms of velocity but, much more convenient is writing it in terms of Mach number. That speed ratio characterizes the velocity in terms of the other relevant speed that is a property of a compressible medium. For the understanding of the motion of compressible media, such as air, it is a requirement to adopt the view that speed is better expressed as Mach number, rather than speed itself. The mathematical and physical descriptions would be excessively complicated. This switch to Mach number as the important flow speed parameter was a key breakthrough by the early aerodynamicists in the 1920s.

An important and practical consequence of the conservation of total enthalpy and temperature is the idea of a *total pressure*. For an isentropic process, the conservation of total enthalpy allows the definition of a pressure resulting from bringing the flow to rest in a gentle, reversible way. The usefulness of that total pressure is in its ability to describe the irreversible (adiabatic, no heat supply or removal) losses associated with a process such a compression in a compressor or flow through a shock wave. Thus, there are total pressure reductions involved in irreversible processes like friction, heat addition, etc., but for a reversible (adiabatic) process total pressure is conserved together with the total temperature.

---

[7] Enthalpy (common symbol h) is the sum of (internal energy, u, *that depends only on temperature*) and flow work (pv). For air, the enthalpy is thus directly proportional to the temperature because air is an *ideal* gas ($pv = RT$).

5.4.1 A Few Details About Specific Heats

This paragraph can be skipped but may be helpful. The intent is to shed light on the origin and nature of the quantity, 'γ', introduced earlier. The internal energy (per unit mass) is proportional to the absolute temperature as is the enthalpy. The proportionality constants for these are called *specific heats*. These quantities are the amount of heat required to raise the temperature of a unit mass of air by one degree. For the internal energy, one can imagine an experiment where one heats the gas in a rigid container to obtain the so-called *specific heat at constant volume*, symbolized by C_v. One can do the same experiment at constant pressure by allowing the test air sample to raise a piston (that keeps the pressure constant) as it is heated. The heat required will be greater because, in addition to raising the temperature of the sample, work is done with a force moving the constraining boundary (force, the *weight* of the piston, times distance upward). The resulting quantity is the *specific heat at constant pressure*, C_p, which is always greater than C_v. From their definitions and that of the enthalpy, h, the difference is the ideal gas constant, R.

For our purposes, both of these specific heats can be taken as constants for typical aerodynamic problems with air. Their numerical values depend on the system of units used (English or metric). Their ratio is, however, dimensionless. In air, the ratio of these two specific heats comes up so often that it is given its own symbol, γ, and is numerically equal to 1.4 for air under normal circumstances of interest to flight. In algebraic relations, the combinations $(\gamma - 1)/2$ and $\gamma/(\gamma - 1)$ appear frequently. Numerically, they are 0.2 and 3.5 respectively but we will keep them in the general form because, well, it looks cool!

So, what have we accomplished so far? We have introduced two abstract quantities that are useful in the description of a gas as it proceeds along a streamline. Looking at the relatively simple flow situation along a streamline, *we find that the sum of static enthalpy and kinetic energy are conserved and temperature is an excellent measure of the enthalpy.* Thus, as velocity increases in the flow over the top of a wing or an airfoil, the (static) temperature measured by a mosquito in that flow element will fall and his environment will warm up again when the flow slows down, while the total enthalpy as characterized by a total temperature remains unchanging.

The second concept is that of the entropy and its close connection to process reversibility. While it cannot be measured by an instrument (nor purchased by the pound), its practical connection is to the total pressure that can be measured. Appendix B sets out a quantitative description of entropy and its relation to measurable descriptive gas properties.

5.5 Total Temperature

The definition of *total temperature* that is often used to describe the heat and work interactions and will be addressed further on in connection with the jet engine. Thus, rewriting the enthalpy conservation statement above in terms of temperature and Mach number, we have:

$$T_{tot} = T\left(1 + \frac{\gamma - 1}{2}M^2\right)$$

For air, the speed of sound, a, is given by

$$a = \sqrt{\gamma R T} \text{ and Mach number, } M = V/a$$

Mach number is a good way to characterize flow speed when dealing with flows that are even a little bit compressible. This statement will become more evident when we talk about the pressure variation along the streamline and the failure of the Bernoulli statement to yield accurate pressures when applied to compressible flows (see discussion at the end of this chapter).

When entropy is conserved, the pressure varies relatively simply with the temperature. As temperature increases so does the pressure and vice versa. This was the last link we needed to be able to relate flow velocity changes to pressure on airfoils and from that deduce forces of interest for flight. The details will be revealed a little further on but, first, a look at a practical manifestation.

To illustrate that the relationship between air temperature and speed we can look at a wing generating lift where we know that the flow proceeds over the wing at high speed. The discussion above suggests that the temperature in the region above the wing should be lower than ambient. If the ambient air is high in humidity, then a decrease in temperature will result in water condensation. Figure 5.1 shows the phenomenon. One might look at this and say that the flow can't be adiabatic because

Fig. 5.1 A Boeing 747 at takeoff on a humid day with condensation cloud above the wing (Photo courtesy Bob Garrard)

condensation is taking place. After all, there is heat involved in condensation. Fortunately, the amount of heat relative to the thermal energy in the air is quite small and we can safely view the flow as close to adiabatic.

With the isentropic relationship between temperature and pressure, the cloud can be viewed as something of a representation of the pressure or lift distribution on the wing. Similar condensation phenomena can sometimes also be seen at the tip vortex of propeller and helicopter rotor blades.

5.6 Bernoulli at Last

We have all the tools required for determining pressures and forces on a wing or airfoil. It may, nevertheless, be convenient to step back and look at the Bernoulli principle because that was the tool early investigators had to analyze aerodynamic issues. Further, the Bernoulli form of the First Law makes for an easier understanding of flow phenomena ... provided the assumption of incompressible fluid is acceptable.

With that limitation in mind, the conservation of total enthalpy for flows of moving water (or low Mach number air) is much simpler. The balance is not between temperature and speed because work, specifically flow work, cannot be done on the fluid because it is incompressible: it cannot change volume. Hence the temperature (as a measure of internal energy) is unchanged as speed changes occur along a streamline. The result is that for incompressible fluids, only the pressure in the flow work term adjusts to speed changes. The energy balance statement that arises is the Bernoulli principle and applies strictly to incompressible fluids like water. For all practical purposes, the enthalpy is a function of only the pressure in this case. The relation between pressure (p) and speed (V) for an incompressible fluid is therefore the familiar statement

$$pv + {}^1\!/_2\, V^2 \text{ or } p + {}^1\!/_2\, \rho V^2 \text{ are conserved along a streamline (the density } \rho = 1/v)$$

In short, as V increases, p must decrease. Further, the forces on a wing consequently vary as the square of the speed.

The failure of this relation to be accurate rests with the fact that, for compressible flows, both pressure and density vary. Therein lies the usefulness of Mach number to characterize compressible flow speeds, rather than using the speed itself. Although it is not as commonly expressed that way, the Bernoulli statement can be expressed (for air) in terms of Mach number as

$$p\left(1 + \frac{\gamma}{2}M^2\right) = p_{tot} \text{ is conserved (incompressible flow).}$$

For the more general statement applied to compressible air, the mathematical statement is somewhat more complicated[8]

---

[8] ... but amounts to the same expression as above when M is small.

$$p\left(1 + \frac{\gamma - 1}{2} M^2\right)^{\gamma/\gamma - 1} = p_{tot}.$$

5.7 Total Pressure

Note in both statements, the notion of *total pressure* is invoked. The total pressure is sometimes referred to as the stagnation pressure, the pressure realized by bringing the flow to rest, nicely. The latter statement follows directly from the definition of total temperature and the relationship between pressure and temperature for an isentropic process.

To put the matter of Bernoulli as an approximation to rest, we note the pressure at the stagnation point of a streamline, namely the pressure realized when the flow comes to rest in air is as shown in Table 5.1 and calculated from the ratio p_{tot} / p for a given Mach number.

The two values are in fair agreement at M ~ 0.3 but differ to a substantial degree at higher Mach numbers.

In closing, one can say that the Bernoulli relationship can be used for a *qualitative* description of the effect of local speed changes on a surface as they impact the local pressure.

Table 5.1 The pressure rise from 1.0 for bringing a flow to rest from a flow at the Mach number noted

Mach no	0.3	0.8	1
Bernoulli	1.061	1.43	1.68
True value	1.064	1.52	1.89

The results are given by the Bernoulli equation as well as by the correct calculation using the energy equation. For M > 1 shock waves are involves and flow description involves more complex aerothermodynamics and using Bernoulli is certainly not a correct approach

Chapter 6
Pressure, Pressure, It's All About Pressure!

With the tools developed so far, the flow around an airfoil can be realistically described by the superposition of the uniform oncoming flow and many small vortices and sources with various strengths distributed along the mean line of airfoil. From a global perspective, the influence of the bound vorticity is to increase the air speed on the top of the airfoil and decease it on the underside. Every part of the vorticity sheet (see Fig. 3.2) has an influence on the flow field and the velocity (and speed) can be calculated everywhere it is desired. The energy equation (including the Bernoulli form) can be used to calculate the local pressure at any point where the velocity is known. Thus, the lifting forces can be determined and all that a wing structure designer might need is available. The pressure distribution from such a calculation is shown in Fig. 6.1. The pressure is above ambient under the airfoil and below ambient on the upper surface. The net lift force and any pitching *torques* (as they called by mechanical engineers) or *moments* (as they are called by aerodynamicists) are the quantities the airplane designer needs to integrate a wing design into that of an airplane design.

6.1 Forces and Moments

The totality of the distribution in Fig. 6.1 is the net lift force. Such a result is from a potential flow analysis consisting of only the uniform on-coming stream and the bound vorticity. The net force is upward. There is zero net force in the freestream direction: no drag. The structure at the physical point on the airfoil where this airfoil is attached (to the walls of a wind tunnel, for example) must deal with that force. For an arbitrary point on the airfoil, the pressure distribution will exert a torque about that point. If that point is near the leading edge, the torque is counterclockwise or nose-down and it is nose-up if near the trailing edge. There is one special point, generally between the leading and trailing edges, where the torque is zero and the whole of the pressure distribution can be summed as just a force. This is the *center*

© The Author(s) 2022
R. Decher, *The Vortex and The Jet*,
https://doi.org/10.1007/978-981-16-8028-1_6

Fig. 6.1 A pictorial representation of the (static) pressure distribution on a typical airfoil at angle of attack. In green, local pressure is lower than ambient pressure; in pink, higher. The net upward lift force on the upper surface (upper blue arrow) is larger and acts on a point to the *rear* of a similar point (lower blue arrow) on the underside which causes the airfoil to exhibit a nose-down pitching moment

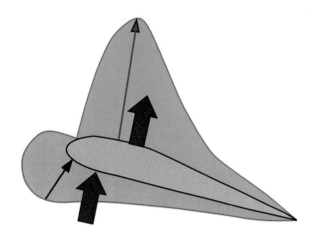

of pressure. Unfortunately for the airplane designer as we shall see, it turns out that the location of the center of pressure changes with changing angle of attack.

In general, both of these reactions (force and torque) are always present in varying amounts, no matter where the mounting point is located. There is, however, one point on the airfoil where the torque is nearly *independent* of the lift force. That point is called the *aerodynamic center* and is located one quarter chord back from the leading edge. This point's location on a wing relative to the airplane's center of gravity is central to the ability of the airplane to fly in level flight, provided that an additional balancing pitching moment, is provided by a horizontal stabilizer surface. In short, for steady, level flight, the designer must balance the inherent pitching moment associated with the airfoil shape, the moment exerted by the weight of the airplane relative to the aerodynamic center, and the moment exerted by the horizontal tail.

The fact that the pitching moment about the aerodynamic center is independent of angle of attack makes size, location, and orientation of the horizontal tail straightforward and thus eases the task of designing the airplane. Parenthetically, we note that the inherent pitching moment introduced by the airfoil shape is a direct measure of its camber. There is no moment about the aerodynamic center for a symmetric airfoil; one without camber.

What about these pitching moments? The pressure distribution on a normally (as shown in Fig. 6.1) cambered airfoil or wing typically results in a nose-down moment about the aerodynamic center. The location of the center of gravity relative to the aerodynamic center also results in nose-down moment. Thus, the moment to be provided by the horizontal tail must be nose-up from a *downward* force behind the wing. Yes, the force on the horizontal tail is downward, against lift. This function of the horizontal tail is difficult to discern in most images or displays because the angles of attack involved are small and orientation of the flow at the tail surface is heavily influenced by the flow behind the wing.

When this negative lift by the tail is not desired, (because it counteracts the wing lift and thus involves a drag increase) a control surface in front of the wing, called a "canard" can provide an upward (lifting) force. The rather uncommon implementation of horizontal control surface in front of the wing in modern aircraft requires artificial (electronic) stabilization because the configuration is not naturally stable in a flying airplane. The word "canard" is French for the bird "duck" and why? You should have asked the person who coined the description watching Alberto Santos-Dumont fly in 1906!

Note the distinction between flight *equilibrium* where all forces and torques are balanced, and *stability* that involves a *return to an equilibrium* for changes in attitude. For a well-designed airplane to fly in steady level flight, equilibrium is required and stability is desirable. With the horizontal control surface in front of the wing, the tendency is for an increase in the angle of attack on the wing to increase further due to the torques and forces so that the pilot has to react and counter this angle of attack increase. In flying their airplanes with forward control surfaces, the Wright brothers were able to do this with their skill. With the horizontal control surface at the rear, changes in condition naturally return to the state before the change so that the flight is stable, i.e., the airplane can be flown "hands off" the controls in the cockpit. The subject of airplane stability is a subject onto itself and not a concern for this development of the forces involved in establishing equilibrium.

6.2 Data for a Simple Airfoil

At this point in the development of the story it might be useful to look at the performance data taken in a wind tunnel test for a simple airfoil. The data is shown in a form that will become clear further on when we take up the subject of dynamic pressure as a quantity for normalizing forces and moments.

A presentation of experimental airfoil data also has to include a notation that portrays the importance of air viscosity. Thus, values of Reynolds number are generally noted. The role of viscous friction and Reynolds number per se is examined in detail further on (Chap. 7), so that the graphs of Fig. 6.2 should be viewed as nominal.

A beautiful compendium of airfoil data is given in a NACA technical report no. 824 dated 1945 (Summary of Airfoil Data) and made widely available by Abbott and von Doenhoff in the classic book on the *Theory of Wing Sections* that every student of aerodynamics encounters. This book is also a good follow-on to this writing with more mathematical details. To illustrate the general nature of the behavior of a simple cambered airfoil and allow the conclusion that the understanding developed so far is quite good, the performance data of a NACA 2412 is shown in Fig. 6.2. The airfoil is similar to the symmetric example (NACA 0012) used in the illustrations later in this story. The appellation of the NACA 2412 is descriptive and means that it cambered by 2% at the 40% chord point and is 12% thick. The percentages quoted are of the chord length. The profile is shown in the second part of the figure. The data given for such a summary of airfoil performance includes lift, drag, and pitching moment about the aerodynamic center as well as the moment about the quarter chord. The

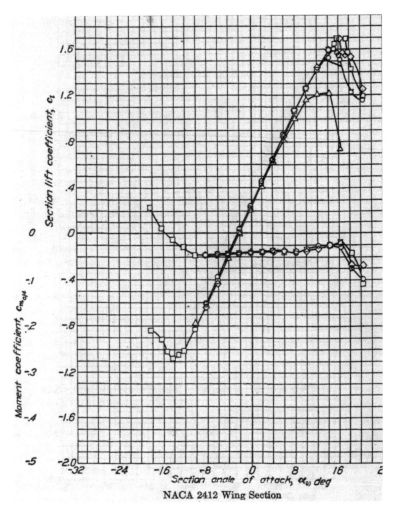

NACA 2412 Wing Section

Fig. 6.2 Performance of a cambered airfoil. The first plot is of lift and moment about the 1/4 chord. The second plot is of drag in in the form of a drag polar and the moment about the aerodynamic center (lower curve). Reynolds numbers (**R**) indicated are in the millions. The reference to standard roughness is to highlight realistic performance contrasting with data obtained with a smooth wind tunnel model (NACA)

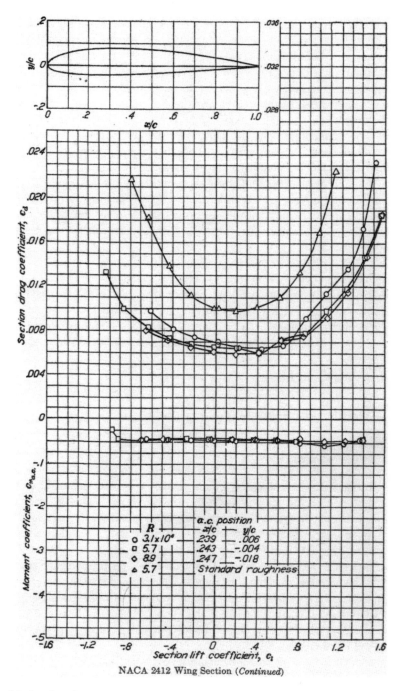

NACA 2412 Wing Section (*Continued*)

Fig. 6.2   (continued)

2412 airfoil illustrates the non-zero nature of the moment about the aerodynamic center for a cambered airfoil. The moment coefficient about the aerodynamic center of the symmetric 0012 airfoil is zero. An examination of the data allows a number of conclusions to be drawn.

Without proof, because we have not developed the mathematics, one can state that the linear slope of the lift curve (lift vs. angle of attack) observed agrees very well with theoretical calculation, i.e., one without consideration of viscous effects. The drag is very small compared to the lift with a maximum value of *lift to drag ratio* on the order of about 150 for a smooth two-dimensional airfoil model. The moment about the aerodynamic center is numerically negative (nose down) and constant for most of the angle of attack range. The aerodynamic center location is given as close to the quarter chord, varying slightly with friction effects. The theory gives the location as exactly the quarter chord. All this is to say that the modeling, in the minds of aerodynamicists and the mathematical one, is very good, especially at modest angles of attack.

The NACA 4-digit series of airfoils is an early (in the history of aviation) design and has been supplemented by many others, especially for drag reduction. The National Advisory Committee on Aeronautics no longer exists as such and current generation of airplane builders largely develop airfoil sections for their special needs using modern computational aerodynamics. The treasure of information gathered by NACA in its early years and summarized in the text above is nevertheless a valuable resource.

6.3 Calculation Iterations Toward Realism

The performance results from wind tunnel tests can be calculated to some degree. The key to doing it successfully involves modelling the behavior of the boundary layer accurately. The pressure distribution on the surfaces of an airfoil allows an examination of the equations of motion for the air in the boundary layer of the airfoil subject to the freestream pressure field. These, to a large degree, allow of the calculation of shear (resistive) forces contributing to drag. The boundary layer alters the effective shape of the airfoil. It is not hard to conclude that accurate determination of these flow phenomena is something of an iterative process involving the given hard surface shape, calculation of the effective shape alteration due to viscous effects and recalculation of the field with the new shape, specifically the changes brought about by air's viscous nature. Probably more than once. Modern analysis of the forces acting on a body like an airplane is not straightforward. It requires the use of powerful computers to be done well.

While the description so far has been limited to two dimensions, a real airplane is a three-dimensional object. The tools available for analysis are similar but they are more demanding because the field is large with three dimensions and three velocity components. Extension of our understanding to three dimensions does not greatly impact the physics of flight through air at a fundamental level. There are, however, important exceptions that involve vorticity. One important aspect of such three-dimensionality is the trailing vortex system.

6.4 Vorticity Provides More Than Lift, It Adds to Drag!

An interesting fact about a vortex is that it cannot end in space. It must end against a boundary or form a closed loop. An example of a vortex ending on a boundary might be a whirlpool in water or a tornado ending on the ground (see Figs. 2.7 or 8.1). Since there are no boundaries in the air through which we fly, the bound vortex on a wing must be a component of a closed loop. Indeed, the wing tips are the front ends of vortices trailing all the way back to a starting vortex left at the airport when the pilot rotated the airplane to begin flight. Together, these elements form a closed vortex "ring". The starting vortex, often illustrated as a line vortex, is actually a sheet (like our spread-out spaghetti bundle) because the airplane cannot be made to develop lift instantaneously.

Figure 6.3 is a view of a much simpler ring. Naturally, over time, the vortex dissipates into randomized turbulence and ultimately to heat when it decays to the molecular level. At that time the vortex will have ceased to exist. More interesting details concerning smoke rings will arise in connection with the nature of the fluid jet, the subject to be taken up in Chap. 9.

The wing tip vortex is, in actuality, a wrapped-up bundle of vorticity shed from the entirety of the span of the wing and not only at the wing tip, especially when flaps are deployed on the wing. This bundle is initially a sheet because the wing, being finite in length, contributes lift to a decreasing degree as one proceeds from the wing root to the wing tip. Figure 5.1 illustrates this idea. The tip generates no lift because

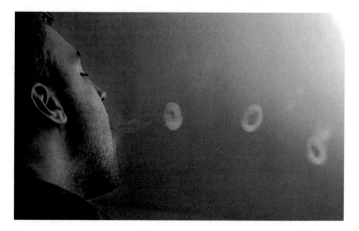

Fig. 6.3 A closed vortex ring with entrapped smoke (Andrew Vargas, Wikimedia, Creative Commons file: http://flickr.com/photo/45665234@N00/2891056110)

the tip allows the upper and lower surface pressures to be equalized by flow from the bottom to the top. As the wing produces *decreasing* amounts of lift toward the tip, it sheds to the rear an *increasing* amount of vorticity which then bundles into a more organized vortex. It may take a few chord lengths for the vortex to appear as a well-defined, symmetric, tornado-like structure.

The flow between the two trailing vortices is downward (see Figs. 2.3 and 6.5), consistent with the downward momentum imparted to the freestream flow as required by Newton's force law. The net effect is that the far field velocity *upstream* is also turned *downward* causing the finite length wing of an airplane to experience flying into a *descending* head wind. This means that the flight lift force vector experienced at the wing is not vertical but is rotated to the rear. In effect, the lift force has a component acting as a drag.

The wing's bound vortex (or better, its distributed vorticity) contributes an upward share to the flow ahead of the wing and adds to the downwash behind it. These velocities play no role on the orientation of the lift force. The vortex cannot interact with itself. The net effect of these influences on the flow ahead of the wing is rather small because the vorticity distributed along the chord and span is modest in length. In a sense, the bound vortex length is shorter than the span of the wing. That makes the relative contribution to the downwash near the wing by the semi-infinite trailing vortex system dominant in its influence.

The penalty associated with force rotation is called *induced drag* because it is induced by (or perhaps better said, associated with) the presence of the trailing vortex system. In practice, the drag can be reduced (though not eliminated) by the use of high aspect ratio wings, such as those of a glider. Roughly stated, the aspect ratio is the ratio of wing-span to chord.[1] A higher aspect ratio wing is slenderer and involves greater skin surface of the wing hence larger skin friction drag. A compromise is always made for a practical airplane design. Elements of the compromise involve speed, fuel consumption, and practical aspects like airport gate access for a commercial airliner.

Figure 6.5 illustrates the use of a conceptual model to aid in understanding the flow configuration. The picture of the trailing vortex shows an approach used in classical texts on aerodynamics to "prove" that one half the of the ultimate downwash velocity appears at the location of the bound vortex and is used to explain the nature of induced drag. The method is well suited for students because the velocity variation along the centerline of the picture can be written in terms of a simple algebraic equation that can be plotted to make the point.

It must be evident, however, that the model fails in a number of respects to illustrate the reality of the situation. For example, the trailing system's induced velocity near the wingtips is certainly incorrect because the downwash velocities associated with the trailing vortices there would be unrealistically large. This is doubly so because the sharp turn of the line vortex is not observed in nature. Finally, it must be recognized that in reality the trailing vortex does not persist as a straight or nearly straight line. The vortex itself is unstable in shape. This matter is not serious for modeling purposes,

---

[1] Technically, Aspect Ratio is (wing-span squared)/wing area. See Appendix C.

however, because the evolution is spread out over space. Figure 6.3 illustrates the phenomenon.

A better model is needed. The tip of the wing cannot support lift. The span-wise lift distribution must be zero at the tip and finite closer to the centerline where the fuselage might be. Just as the vorticity is distributed along the chord, it is also distributed along the span. As one looks from airplane centerline to wing tip, the decreasing lift must shed an increasing amount of vorticity into the stream as a sheet. Such a model softens the singularity at the wing tip. Yet, we observe that downstream of the wing the vorticity does bundle into a fairly well-organized line vortex (see Fig. 6.4). Adaptive modeling is again called for. A better model is shown in Fig. 6.6. The model involves a multitude of smaller loops of vorticity better reflecting the span-wise lift distribution. The loops are not shown in the figure (because the figure would be too complex) but the span-wise variation of the shed vorticity is shown as region D.

In classical aerodynamics, the variation of the vorticity of the trailing system shown in Fig. 6.6 is used to determine the lift distribution to give minimum induced drag. That variation turns out to be an elliptical one, meaning that when plotted along the span, the line representing lift is an ellipse (region A). That condition is attained when the downwash at the wing is uniform (shown as region C in the figure). That result might be an expected one for any other distribution for a given lift requires a greater investment in energy to bring it about. Such a model analysis was the basis for an elliptic wing planform of the Supermarine Spitfire designed and used in WW II. In practice, tapered wings are easier to build and perform nearly as well.

While Fig. 6.6 shows a rectangular wing (B), the lift distribution may be made elliptical by twisting the wing and forcing the wing to operate with a varying distribution of angles of attack.

The model that leads to the conclusions about uniform downwash assumes, for analysis purposes, that the vortex sheet extends far downstream unchanged. That

Fig. 6.4 Left: A fighter plane with trailing vortices made visible by the condensation of moisture in the air. Near the center of the vortex core the high velocity leads to locally low temperatures and condensation (US Air Force) photo. Right: The vortex nature of the flow is well illustrated with smoke injection at the wing tip on a test airplane. The airplane is a Boeing 727. The smoke injection ports are visible in Fig. 2.2. (NASA Dryden Research Center)

eases the mathematics but is not an accurate reflection of reality because the sheet bundles into a line vortex far downstream as described above. Perhaps the use of the spaghetti noodle bundle is appropriate again here, but the noodles would have to have been cooked to twist into a line vortex! Approximate models are, nevertheless, useful to gain insight into the flow kinematics and to obtain some degree of certainty about the magnitude of the effects involved on what is really wanted: a good measure of the actual induced drag associated with a specific design.

The simple models outlined in Figs. 6.5 and 6.6 are simplifications of the real world. Modern computers allow much more detailed representations of the flow around a winged airplane. The improvements allow incorporation of span-wise and chord-wise lift distributions. Also available for inclusion are better descriptions of the flow field associated with the coalescence of the vorticity sheet from the wing into a trailing line vortex pair that exert influence on one another.

The modeling exposed in this discussion and that of the related propeller discussion in Chap. 9 owes a lot of the original thinking to Ludwig Prandtl, the German

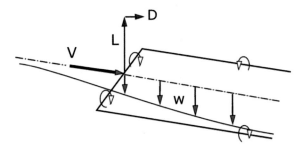

Fig. 6.5 Simplified sketch illustrating induced drag at mid span from the rotation of the oncoming flow vector by the wing's trailing vortex system. On the wing-bound vortex, the downwash on the centerline has half the value experienced at the far right. The oncoming stream (V) is therewith rotated downward (green arrowhead) and the lift force vector (L) is rotated to the rear, contributing to drag (D). The vectors (oncoming air speed (**V** + **w**) and the resulting force, **L** + **D**) are 90° apart. Far to the right (at the airport) would be a "starting vortex" closing the loop

Fig. 6.6 A more realistic model of trailing vortex system (D) from a wing (B) with distributed lift. Noted are the lift (A) and downwash (C) distributions. In the far-field to the right, the distributed vorticity (D) shed as a sheet bundles into an organized line (trailing) vortex further downstream

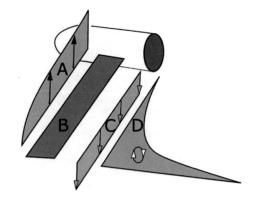

aerodynamicist who did much of the pioneering work in the early in the twentieth century. His understanding was furthered by Hermann Glauert, the British aerodynamicist. Both individuals left their marks on the field with the latter well known to aerodynamics students for his book *"The Elements of Aerofoil and Airscrew Theory."* It was, for a time, required reading, including for this student. A simple summary of a quantified description of induced drag is given in Appendix C.

6.5 Flight in Ground Effect

The model of a bound vortex with trailing vortices can be used effectively to describe flight in so-called *ground effect*. The details are left for the serious student, but the essence is that the ground in proximity of a flying wing is a flat, rigid surface. Aerodynamically such a rigid surface is obtained with a second bound and trailing vortex system on the other side of the surface. That way an impenetrable surface results from the addition of the two vortex systems. The system "below the ground", so to speak, induces a velocity field that reduces induced drag. A pilot in an airplane approaching the ground for landing typically feels the reduction of drag as a small net forward acceleration.

6.6 Dynamic Pressure and Those Convenient and Pesky Coefficients

When we look at the pressures that sum up to lift, we are concerned with only that part which results from the motion of the air. When flight speeds were slow, the Bernoulli form of what we now call a flow energy equation was simple. It involved the dynamic nature of motion in a single term, the *dynamic pressure*, $1/2 \, V^2 1/2 \, \rho V^2$ given its own symbol, q.

Since the pressure differences from atmospheric pressure are of primary interest, one would expect that all pressures on an airfoil can be related to the dynamic pressure. This indeed the case. Thus, to simplify life, aerodynamicists express all pressures from any calculation related to airfoils in a non-dimensional way as a *pressure coefficient*[2] written as

$$C_p = \left(p - p_{in\,the\,freestream\,far\,away}\right)/q$$

This coefficient is the quantity displayed in Fig. 6.1: negative above the airfoil and positive below.

The following question may arise in the mind of the reader: "Does the presence of a boundary layer cause a pressure measurement made on a surface to differ from the

---

[2] The symbol C_p is the same as the specific heat but the context makes distinction evident.

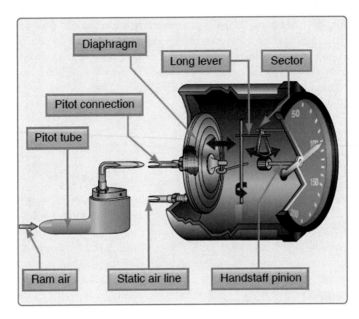

Fig. 6.7 Cross-section of a bellows type Indicated Air Speed instrument connected to a pitot probe (image courtesy of *Flight Literacy*)

pressure calculated from consideration of the inviscid freestream flow field?" The answer is that the difference is extremely small. Figures 6.1 and, for example, 7.2 in the next chapter rely on the premise that the pressures are identical. The argument for that truth is made in Appendix B and involves the invocation of Newton's second law of motion.

6.7   The Pitot Probe and Dynamic Pressure

There are two operational aspects of the dynamic pressure of interest to the pilot of an airplane. The first is that dynamic pressure plays an important role in determining the loads experienced by the surfaces of an airplane. These cannot be allowed to exceed the design values lest a structural failure be risked. The second is that dynamic pressure has to be large enough for flight to be possible. Stated another way, one could ask: When can the airplane accelerating down a runway be rotated to present a positive angle of attack on the wings and thereby lift off the airplane the ground?

Dynamic pressure is relatively easy to measure and is the basis for telling a pilot of his flight situation, the airplane's speed in particular and, more importantly, whether the wing can operate effectively. The determination of dynamic pressure is typically

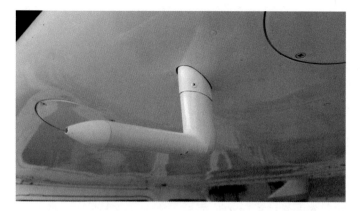

Fig. 6.8 Pitot probe under the leading edge of a Cessna 172. The ports for measuring static pressure are on the airplane fuselage (Photo: Ronald Parker)

made with an instrument (Fig. 6.7) consisting of a bellows[3] wherein stagnation pressure air is fed to its inside and the outside is exposed to static pressure. The deflection of the diaphragm responding to the pressure *difference* allows a connected needle to display dynamic pressure on a dial. At modest speeds, the difference between the total (also called stagnation or ram) pressure and the static pressure is the correct value of the dynamic pressure. In very old airplane instruments, the information might have been displayed as deflection of a liquid in a glass U-tube where one end of the "U" is connected to the pitot probe and the other is left open, exposed to the nearly static pressure in the cockpit. That turned out to be awkward in a number of ways in that the pressure is hard to read, the fluid could be spilled (when flying upside down), and the reading depends on the g-forces experienced. Dial instruments were quickly developed and adopted. Ever more commonly today, the pressure signals are fed to transducers that generate an electrical signal used to display the information on an electrical instrument.

The stagnation pressure is measured by a probe like that shown in Fig. 6.7, also shown on Fig. 6.8. It brings the flow to rest on the relatively blunt nose. For the best measurement, the probe should be positioned at a location where it is exposed to clean flow. Locating it at a place where it faces true freestream pressure is often impractical, however, because the pressure field around an airplane is large, reaching out distances of many feet. The underwing location of the probe on a general aviation airplane in Fig. 6.8 is a reasonably good location. On commercial airliners, the probe is typically located near the nose of the airplane.

Similarly, the static pressure should be measured far from the airplane so that its pressure field associated with and surrounding the airplane is not involved in the measurement. Again, in a practical setting, a local measurement can be made on the probe itself by means of ports (small holes) near the elbow of a pitot probe

---

[3] A bellows is a set of two flexible diaphragms that separate when the pressure inside is raised relative to outside.

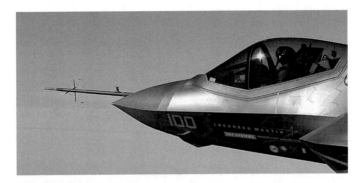

Fig. 6.9 Pitot tube on a Lockheed Martin F-35. This is typically a temporary installation for flight test purposes. This rig measures a number of aerodynamic and attitude parameters. The static measurement is made along the long probe even when flight speed is supersonic [Photo: Wikimedia Commons File:CF-1 flight test.jpg (cropped)]

or it is taken from a static pressure port located somewhere on the airplane. Any inaccuracy in both these measurements is either ignored (because it is negligibly small) or accounted for by a calibration. Such a calibration is usually established with prototype test aircraft with a long nose probe (Fig. 6.9) or on a long, thin trailing tube behind the test airplane. In either case, the goal of the calibration is to measure the true pressure of the environment as if the airplane was not present and relate that measurement to the pressure actually measured at the locations deemed practical. The calibrations are then built into the display of pressure by adjustments on the instrument or mathematically for display in digital form.

Dynamic pressure is important for assuring that the wing can generate the necessary lift force while staying within the structural load limits. It may be measured in pounds per square inch, inches (or millimeters) of mercury, Newtons per square meter, or other units. The call-out of such abstract quantities are not convenient for the pilot. He or she deals with speed. Fortunately, a speed can be inferred or calculated from the dynamic pressure provided the air density is known. An easy solution is to build into the pressure measuring gage the density of standard sea-level air and display the pressure as a speed by marking the dial appropriately. Traditionally, the speed determined this way is measured in knots (nautical miles per hour, reminding us of a tradition established by mariners).

By these means, the pilot can think of dynamic pressure as a speed. The instrument that displays the dynamic pressure is an *indicated airspeed* indicator (Fig. 6.10). The numerical value shown on the gage is that of the airplane as if it was flying at sea level on a standard day. When flight is at higher altitudes, the indicated speed shown on the instrument will be smaller than the *true airspeed* whose accurate determination would have to include data about the local air density where the airplane is flying (or trying to fly!). This value of the local density can be determined from measurements of temperature and pressure of the airplanes' environment with application of the ideal gas law ($pv = RT$) since density is $1/v = p/RT$. Such measurements are quite

Fig. 6.10 Simple display in an airplane instrument panel showing indicated air speed. In the U.S., this speed is traditionally given in knots (nautical miles per hour). The green zone is safe operation range, yellow requires smooth air, and red tab is the not-to-exceed speed for concerns of structural damage to the airplane. The white range is for operation with flaps, such as during takeoff or landing. More complex versions of such a dial are common (Wikimedia Commons: Marek Cel, file: Airspeed indicator.svg)

commonly made on aircraft for many purposes so that true air speed may be obtained from proper instruments or a hand calculation.

A good illustration of the role of dynamic pressure is in connection with takeoff from an airport. Independent of whether the airport is a high altitude or sea level, the indicated speeds necessary for the pilot to rotate the airplane for lift off are identical. Indication on the gage in the lowest reading in the green zone (Fig. 6.10) is where the pilot can initiate rotation. With flaps lowered, he or she can do it earlier in the takeoff roll (the white zone). The ground rolling speed is higher for the mountain runway takeoff. At the 130 knots mark, the pressure measured is on the order of a half a pound per square inch, one inch of mercury, or 0.035 atmospheres. A commercial airliner in cruise would experience about 4 times this dynamic pressure or double an indicated speed, around 280 knots with a maximum of about 350 knots. The upper end of the white zone is where flaps must be retracted because there is a risk of their failing to stay with the airplane due to excessive structural loads. The yellow and finally red lines display other structurally imposed speed limits on the airplane.

The pilot would also want to know how fast he is going relative to the ground so that he can navigate accurately. Within the airplane, however, the best speed determination is that of *true air speed*, the speed of progress within the air mass around the airplane. Wind will inject a difference between the true and ground speeds. Nowadays ground speed is often obtained from GPS (Global Positioning Satellite) information obtained as input to navigation systems. From such systems, wind information is easily deduced. GPS systems are common in general aviation aircraft and are uniformly installed in all commercial aircraft.

The pitot probe and the associated indicated air speed indicator remain a fixture on most subsonic aircraft such as general aviation aircraft and commercial airliners. No matter how sophisticated the pressure measurements may be made and displayed, flight safety consideration usually require the pitot probe and its instrument as a backup resource.

So far, the discussion of flight speed(s) has not addressed what is the best speed for an airplane to travel? Maximum speeds are always interesting from a general viewpoint, but operators of airplanes usually consider economic issues as a more important concern. The achievement of long range with minimal use of fuel is paramount for commercial airline operators and the designers of the airplanes. General aviation flyers are often similarly concerned with the fuel consumption and the issue of the related cost. The speeds that allow addressing this goal is where the discussion to follow will have to take us.

Chapter 7
Putting It All Together

If it were not for friction and compressibility, the analysis of flow phenomena would be pretty simple. Friction is a necessary ingredient for lift and it controls most of the price we pay in propelling an airplane through air. Knowledge of drag as a resistive force acting on airplane surfaces is important in allowing the efficiency of the airfoil or wing to be determined and has important consequences for the performance of the airplane. Compressibility is a matter we must deal with—if we wish to fly fast.

So, how do engineers describe the performance of airplane or their wings? The first requirement is to make the presentation of such information as simple, universal, and as compact as possible.

Since the lift force scales with dynamic pressure (q),[1] it is customarily expressed as a *lift coefficient*, that is proportional to lift. Because the lift force is the result of pressure acting over an area, the lift coefficient is written in terms of the only area in the problem, namely the chord length of the two-dimensional airfoil or the wing area for a three-dimensional wing. Thus, we have for a two-dimensional airfoil or a wing

$$\text{Section lift coefficient} = c_l = \text{Lift}/(q)(\text{chord}) \text{ or}$$
$$\text{Wing lift coefficient} = C_L = \text{Lift}/(q)(\text{wing area})$$

There are a similar definitions for airfoil, wing or airplane drag coefficient, C_D. Similarly, the section moment coefficients shown in Fig. 6.2 are defined by

$$\text{Moment coefficient} = c_{mz} = \text{Moment at } z/(q)(\text{chord})^2$$

The moment must be specified as measured at a specific point which is usually the aerodynamic center or the quarter chord (see Fig. 6.2). For a wing or an airplane, the normalizing parameter is the product of wing area and the mean aerodynamic chord.

---

[1] See cautionary note in the introduction about symbols with multiple meanings.

© The Author(s) 2022
R. Decher, *The Vortex and The Jet*,
https://doi.org/10.1007/978-981-16-8028-1_7

Fig. 7.1 Representative lift curve for an airfoil in incompressible flow ($M \sim$ 0.2) and compressible (subsonic, M indicated) flow. Note the departure from linearity and greater steepness at the highest Mach number

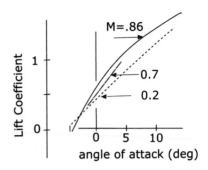

These are the section coefficients displayed in Fig. 6.2 for air at low Mach number, i.e., for incompressible flow. Let's examine the role compressibility plays in determining lift coefficient. The variation of this measure of lift with angle of attack is what the inviscid theory with incompressible air correctly predicts for small angles. An example for a wing (C_L) tested in a wind tunnel at various Mach numbers is shown in Fig. 7.1. This particular airfoil is cambered because it provides lift at zero angle of attack. The Mach number plot line tagged with $M = 0.2$ is a good approximation of incompressible flow. That line agrees very well with calculations we might make using the analysis tools developed so far. Extension of such a plot to include compressibility effects involves the higher freestream Mach numbers as noted.

A couple of observations are noteworthy. The maximum lift coefficient is of order 1.0 and could be larger, especially for airfoils designed with high lift devices. The slopes of the various curves at transonic Mach numbers differ somewhat because the flow is compressible. From this it can be concluded that lift is only modestly affected by compressibility, when we limit our attention to the angle of attack range and transonic flight speeds where airplanes like commercial airliners operate.

7.1   Wing Lift Performance with Viscous Air

What about the lift performance of wings in air with a greater role played by friction effects? Fig. 7.2 is the pressure distribution one might find on a simple 12% thick symmetric airfoil (NACA 0012) at 10 degrees angle of attack. The leading edge is at $x/c = 0$ and the trailing edge is at $x/c = 1$. The Mach number is low enough for the flow to be incompressible. The pressure is measured by ports (small holes) in the airfoil each connected to a pressure sensing instrument. At the various high Reynolds numbers, the role of viscosity (or friction effects) should be modest. The experimental data displays fairly consistent flow behavior even as various means are employed to examine the role of boundary layers allowed to transition to turbulent on their own (free transition) or forced to do so by a layer of roughness on the airfoil (fixed transition). This aspect is a detail that is beyond the level of this discussion. Note the very high flow acceleration rate around the leading edge to the upper surface.

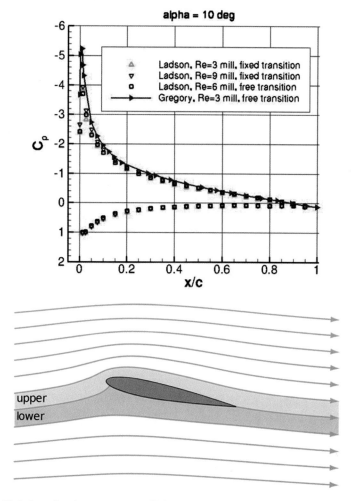

Fig. 7.2 Variation of surface pressure coefficient along the chord for a symmetric (uncambered) airfoil (NACA 0012) at high Reynolds numbers indicated and with various boundary layer transitions to turbulent flow employed. The airfoil shape is shown at the bottom of the figure. Note the stagnation point on the lower surface where $C_p = 1.0$. Note also the increasing pressure on aft portion of the upper surface (falling C_p) behind the point of minimum pressure (data: NASA Langley Research Center; airfoil image: Michael Belisle: Wikimedia Commons, file: Streamlines around a NACA 0012-hu.svg)

The use of the coefficient to represent pressures also coalesces all the data for various speeds at which the wind tunnel was run to nearly identical curves, simplifying the engineers' work.

When Reynolds number is under half a million, we might expect the flow to be primarily laminar. The airfoil performance at modest Reynolds numbers is illustrated in Fig. 7.3. Under such conditions we have fairly consistent lift behavior at small

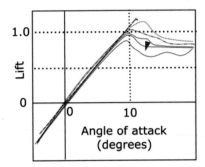

Fig. 7.3 Variation of lift (as a lift coefficient) with angle of attack and modest Reynolds number based on chord. The data is experimental for the NACA 0012 airfoil. The dashed red line is to indicate the theoretical line to be expected with inviscid flow. The arrow pointing downward is in the direction of decreasing Reynolds number. The values of *Re* (in thousands) varies from 290 (top curve), 200, 120, 80 and 50 at the bottom (Data adapted from Researchgate.net)

angles of attack that agrees well with an inviscid flow analysis. At higher angles of attack, the situation worsens because the lift curve no longer follows the linear portion of the curve.

7.2 Stall

When an airfoil's angle of attack is increased beyond the linear portion of the lift curve (Fig. 7.1), the lift will eventually fail to increase and then decrease. The departure of such a curve from linearity is called *stall*. At the lower Reynolds numbers shown in Fig. 7.3, stall occurs around 10–12°. At Reynolds numbers in the millions, with a lesser importance of friction, the stall angle of attack for the same airfoil turns out to be about 16°. The data for NACA 0012 is not presented but similar trends are noted for NACA 2412 shown in Fig. 6.2). At high angles of attack, one should expect deleterious consequences not only for drag but also on lift characteristics.

In practice, flight is normally at lower angles, but it is important to know how the wing behaves near stall.

7.3 Why Does a Wing Stall?

The boundary layer is important for its role in influencing the behavior of the freestream flow. Consider that by shaping a surface to generate a lifting force, we are imposing boundaries to influence the direction and speed of the air. We try to insist that the flow accelerate to high speed (for low pressure, see Fig. 6.1 and 7.2, low negative C_p). We then motivate the flow to decelerate again to reach a pressure close to that of the original incoming flow at the trailing edge. This is evident in Fig. 7.2 because C_p is near zero at the trailing edge where it is, in fact, difficult to measure because of the local thinness. It is this deceleration to the trailing edge that is the principal challenge to the design and operation of wings and wing-like structures such as, for example, propellers or compressor blades in a jet engine.

Note that, even when operating nicely, the measured pressure coefficient is positive near the trailing edge. This means that the nearly stagnant air slowed by the boundary layers on both sides of the airfoil allows for transmission of the high-pressure information from the bottom to be shared with flow over the upper surface to about the 90% chord length point in this example.

7.4 Adverse Pressure

The air in the boundary layer along the upper surface of an airfoil is challenged by the reality that it does not have the energy required to rise back up to the high pressure that the inviscid freestream air outside this layer can. Friction has robbed it of momentum and therefore also of energy. The increasing pressure in the flow direction is an *adverse pressure gradient*; adverse meaning unfavorable. Air would prefer to flow into a region of *lower* pressure.

The freestream air can handle this change because it has sufficient kinetic energy that it can convert back to thermal energy (technically stated as enthalpy, which in incompressible flow leads to a rise in pressure). The boundary layer air cannot. When the point is reached where all kinetic energy in the boundary layer air is used up, the flow is stationary. With the pressure ahead still higher, the natural reaction is that the stationary air will simply flow backwards, i.e., upstream. With that, the flow fails to follow the prescribed flow direction and is said to *separate* from the boundary prescribing the flow geometry. At even larger angles of attack, useful lift is lost entirely.

This is the mechanism that typically leads to wing stall. It is also the mechanism that determines the lift and drag of the wing at all angles of attack. The flow separation near the trailing edge is central to the drag experienced. As stall becomes more pronounced, the separation point moves forward on the wing resulting in a wider wake. Figure 7.4 illustrates the relevant flow phenomena near the onset of stall. Figure 6.2 shows the general form of the variation of drag of a simple airfoil (NACA 2412) at various angles of attack (at low Mach number: incompressible flow). The camber on this airfoil leads to a minimum drag at a non-zero lifting condition, which is desirable for an airplane. The drag minimum is rather broad over a range of lift but, increases rapidly for high angles of attack.

Stall occurring at the trailing edge is the common mechanism for wings or airfoil at high angles of attack. It is, however, not the only mechanism. The leading edge can also be the culprit with the formation of a bubble of separated flow near the point of minimum pressure. Good design of the leading-edge geometry, the use of leading-edge flaps or slats, and, as we shall see later, the use of devices designed to

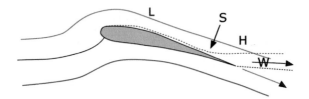

Fig. 7.4 An airfoil at an angle of attack sufficient to cause flow separation at the point noted (S). Regions of low (L) and high (H) pressures are noted. The dotted line is the boundary layer edge. The lift is also reduced by virtue of the downward flow deflection [black arrow showing direction of the wake (W)] not being as demanded by the trailing edge direction (green arrow). The green streamlines shown would be as shown without flow separation

forestall (pun intended) stall (called vortex generators) are very effective in making stall a manageable aspect of operating a wing for high lift.

7.5 Transonic Flight

Modern airliners cruise at speeds, called transonic, because the flow over the upper surface of the wing is locally supersonic. In this flight regime, there is a bubble of supersonic, low pressure, flow that must return to the subsonic, higher-pressure region near and aft of the wing's trailing edge. In this supersonic "bubble", and outside the boundary layer, no information can travel upstream from the trailing edge to tell the flow to slow down because the flow speed is faster than the speed with which information can travel.

The only way the matter can be reconciled is with a wave that suddenly changes to supersonic flow to subsonic. This is a *shock wave*. Across this wave, all descriptive parameters of the flow change: velocity, pressure, temperature, density, and speed of sound. The velocity decreases while all the others increase. The flow energy is conserved so that the total temperature is conserved, but since this situation cannot be reversible, entropy increases and consequently, the total pressure decreases. These changes are modest when the Mach number is above and close to one and they become more substantial when the flow Mach number reached at the front of the shock wave is larger.

That entropy should not be conserved though a shock wave is to be expected since the reverse situation of a low-speed flow suddenly accelerating to higher speed for no good reason is not likely (not ever) to be observed. Flow through a shock wave is not reversible.

Flight speeds of commercial jets are around 0.8–0.85 Mach number. A passenger sitting over the wing might, under the right circumstances, be able to see the stationary shock wave on the wing because the density change across the shock will result in a change of the refractive index of air. The change is quite small and could be visible when we look toward the horizon. However, under most circumstances, the sky on

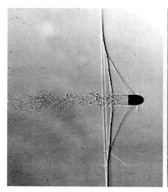 

Fig. 7.5 Left: A shadowgraph of a bullet in supersonic flight. Weak oblique waves from the nose coalesce into an oblique shock that turns into a normal shock further out. Note the turbulence in the wake (Photo: Daniel P.B. Smith, Creative Commons, 1962; https://en.wikipedia.org/wiki/Doc_Edgerton#/media/File:Shockwave.jpg). Right: Shock image generated by a low-level fly-by of a US Navy "Blue Angels" F/A-18. Note the wing recompression shock above and below the wing trailing edges (above and below the number "5") as well as shocks at the leading edge and above the cockpit. Its visibility is due to the refraction from the air density change across the shock. There is also a faint amount of condensation above the low-pressure, upper side of the wing (Image courtesy Barry Latter)

the horizon has no structure against which optical distortion could be detected by the human eye. We have to "see" the shock by other means. For example, on a flight at low latitudes or an east–west flight near noon with the sun directly overhead, the sun's rays will also be refracted by the shock because of the abrupt change in density. Such refraction will result in a shadow on the wing surface. When visible, this shadow will typically dance fore and aft as the atmospheric conditions of the flight oscillate ever so slightly. The physics of this phenomenon is utilized in the generation of a *shadowgraph* for the purpose of flow visualization. Figure 7.5 illustrates the technique with a photo of a supersonic bullet with its shock wave system and the wake that follows it.

To illustrate the optics of a shock wave in circumstances that might be more readily observable, the right image in Fig. 7.5 is of an airplane flying close to the speed of sound past nearby houses. These provide a good background against which the waves become visible. The US Navy's Blue Angels fly-by over Lake Washington waters in Seattle was entertaining and very loud - for a very short time! Note also the recompression shock aft of and above the cockpit.

Today's commercial airliners routinely fly with supersonic flow on the upper surface of the wing but only up to the speed where the waves are weak and don't lead to a significant drag rise, see Fig. 7.6. The sketch (Fig. 7.7) illustrates the flow separation with a shock wave strong enough to cause flow separation and an associated drag increase. The separation is caused by the shock establishing regions of low and high pressure that are easily connected by a segment of the boundary layer air that allows flow upstream, from high pressure behind the wave to low pressure

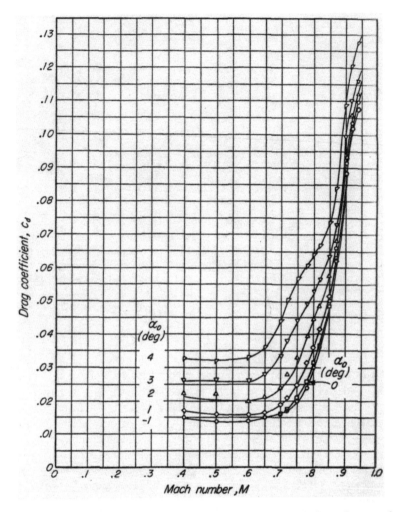

Fig. 7.6 Drag behavior of a NACA 2312 airfoil at speed near the speed of sound at a number of angles of attack. The power required to overcome drag varies as M^3 which implies a sharper rise in power required than shown in this plot (NACA)

ahead of it. Flow separation occurs if the wave is strong enough, a situation generally avoided in commercial aviation. The subject of shock-boundary layer interaction has been a matter of intense study by government, academic research laboratories, and airplane builders for many decades. It plays an especially strong role in determining the performance of inlets on high-speed military jets. On wings, lift will also be affected by separation as suggested in Figs. 7.4 and 7.7.

An airfoil drag test at various Mach numbers approaching $M = 1$ is shown in Fig. 7.6. While this and the other airfoils described in this text, are not likely to be suitable in practice, their characteristics are representative of all airfoils in a

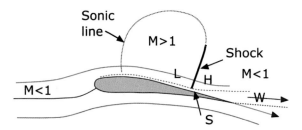

Fig. 7.7 Flow separation (point "S") due to the recompression shock on a wing at transonic speed. The low-pressure region on the upper surface is locally supersonic with a shock strong enough to cause the boundary layer to detach from the surface. The creation of the wake (W) is the source of increased drag. L and H refer to the static pressures near the shock wave and imposed on the boundary

qualitative, rather than quantitative, sense. The steep drag rise above $M \sim 0.6$ (for this airfoil) is termed the transonic drag rise. Realistic airfoil designs for airliners, for example, have a significantly higher drag rise Mach numbers.

The rapid wing and airplane drag rise with increasing speed associated with the effects of shock waves and shock/boundary layer interaction was, for a time, a difficult barrier for flying at speeds close to that of sound. In the days during and after WW II when serious attempts at high-speed flight were pondered, the power available from the piston engines of the day was insufficient for the propeller to provide the thrust required. A second and more important reason for the difficulty of reaching near sonic airplane flight speeds is that the propeller blade always deals with air at relative speeds greater than the flight speed. The turning blade sees oncoming air as the vector sum of flight and blade rotation speeds. Consequently, it is the propeller blade that encounters a serious drag rise before the airplane does as it tries to increase speed. The prospects for flying faster looked better with the new jet engines. The first supersonic flight was, however, achieved by a rocket powered airplane. The jet engine was not yet ready.

The people making early attempts at flying very fast termed the drag increase associated with high-speed flight (Fig. 7.6) as running into the *sound barrier*. It is not a barrier as such and holds the flyer back from flying faster only if the power or thrust necessary to overcome the drag is not available. It turns out that on the other side of the sound barrier, further from Mach one, things get a little easier and flight in the regime is now commonplace—for military airplanes.

7.6   Shock Waves

Just what is a shock wave? It is the sudden, albeit partial, conversion of organized flow kinetic energy to disorganized thermal energy at the molecular level. Such a process cannot occur in reverse and hence is irreversible.

The sudden conversion of kinetic energy to another energy form is a possibility whenever a mechanism exists that can accept the converted energy. An interesting example closer to everyday experience is the hydraulic jump where the energy absorbing mechanism is potential energy by a liquid medium, nominally water, in a gravity field. The similarity between shock waves and hydraulic jumps is quite astounding and interesting. In both cases, the flow velocity ahead of the jump is faster than the relevant wave propagation velocity. In the case of hydraulics, the propagation is by *surface waves* (not compression sound waves) for which the relevant speed is proportional to the depth of the surface of a stream. In a sense, liquid depths and gas pressure play similar descriptive roles. A simple experiment in the kitchen sink (Fig. 7.8) is illustrative of the "shock" wave phenomenon. The flow ahead of the "jump" is supercritical (in the language of hydraulics, the equivalent to supersonic for us) and the flow behind the wave is subcritical. This assertion can be tested by seeing whether the flow does or does not support steady waves: a standing wave is supported by a slim object held into the water ahead of the jump, but not

Fig. 7.8 A kitchen sink water flow with a hydraulic jump. The flow away from the impact point can be described as a source flow (Photo: Wikimedia Commons: James Kilfiger, File:Hydraulic jump in sink.jpg)

behind it. The flow ahead of the jump is shallow and rapid while is slow and deep behind it. Similar flow phenomena can be seen in a stream flowing between rocks or at the end of a spillway chute of a dam.

7.7   Condensation at High Speed

This book is primarily about subsonic aerodynamics but high-speed flight invariably involves supersonic flow. The next few paragraphs focus on that aspect of aerodynamics because it is necessary to explain some interesting observed phenomena. Figure 5.1 shows condensation of water vapor above the wing of an airplane at modest speed. Similar images are available on airplanes at transonic speeds and are often used as illustrations of the shock wave system on the wing of an airplane under these flight conditions. For example, Fig. 7.9 shows an F/A-18 flying in air that is close to saturated with water vapor. Evident is a cloud of vapor that follows the airplane. The viewer should not interpret this as visualization of the shock wave system associated with the wing and airplane. Clarification of this phenomenon is an opportunity to delve into the nature of shock and other waves on wings that play a role here but

Fig. 7.9   Photo of a F/A-18 in transonic flight in saturated air. The text outlines the argument that the cloud boundary is not a representation of the invisible shock wave system (Photo credit: https:// commons.wikimedia.org/w/index.php?curid=15250934)

are not visible. This introduction of the topic will also serve to aid the discussion of flows into inlets and from nozzles.

7.8   Supersonic Wings

The flow physics at $M = 1$ is rather complicated because parts of the field are subsonic while others are supersonic. We can make a few simplifying assumptions to better understand what is seen in Fig. 7.9. First, we note that the F/A-18 airplane has a wing that is swept to a very small degree so that looking at it as a straight the wing in two dimensions is appropriate. The wing can be modeled as a flat plate because it is also rather thin. Second, the aircraft in the figure is flying at a Mach number that is not specified but probably rather close to $M = 1$. Another useful assumption is to take the flight Mach number to be somewhat larger so that the behavior of waves can be illustrated more clearly.

In mildly supersonic flight, say at Mach number of 1.4, the wing creates a wave pattern that is rather simple. At supersonic speeds, the wing cannot transmit information about its presence to the flow ahead of the wing. A wave is the sudden manifestation of the interaction between the wing and its environment. This contrasts starkly with the situation when the flow is subsonic and adjusts gently to the geometry demanded by the airfoil. At a positive (lifting) angle of attack, the leading edge of a supersonic flow generates two kinds of waves. The flow under the wing is turned into itself and is compressed. This is similar to the physics that affixes bound vorticity to the airfoil with incompressible flow as illustrated in Fig. 2.5. The wave is a compression or shock wave. When the turning angles are small, and the wave is referred to as an oblique wave. Shock waves are sometimes *normal* shock waves that are stronger and lead to a greater total pressure loss.

The wave encountered by the flow above the wing causes the flow to occupy a greater space and is therefore expanded. This expansion wave is not sudden but spread over a region through such an angle that the flow follows the direction imposed by the upper surface. Such a wave is called an expansion (or Prandtl-Meyer) fan. The flow through it is gentle enough to be reversible.

Figure 7.10 illustrates the wave pattern when a very thin flat plate is set at an angle of attack in supersonic flow. Because the pressure above the wing is lower than ambient pressure and higher below the wing, it generates lift. Because the flow velocity is uniform along both surfaces, the center of pressure is at the half chord point and firmly located there, irrespective of small changes in angle of attack. In this fact resides the vexing challenge of the early flyers attempting to fly faster than the speed of sound, because the airplane control mechanics change in character at transonic speeds.

At the flat plate airfoil's trailing edge is another set of waves. The upper surface flow is slowed with its locally higher (than freestream or lower surface) Mach number. A compression wave allows departure from the trailing edge parallel the lower surface flow. The lower surface flow undergoes an expansion wave. The two shock waves

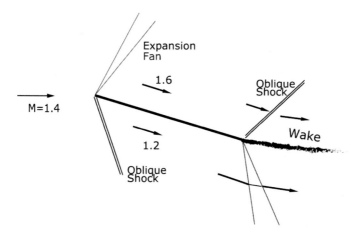

Fig. 7.10 The wave pattern on a lifting flat plate airfoil in supersonic flow. Note the two kinds of waves: oblique shocks and expansion fans. The values of approximate local Mach number are noted

(lower leading edge and rear trailing edge), having dealt with differing Mach numbers in their oncoming flows are different in strengths so that the entropy levels of the flows on either side of the wake differ. The trailing edge shock wave is stronger than the leading edge one. This difference in entropy is manifest in the existence of a wake. Its orientation is in a downward direction, consistent with Newton's laws.

A real wing is not a flat plate but would consist of a sharp leading (w)edge and curved upper and lower surfaces to close at the trailing edge. Figure 7.11 illustrates the (exaggerated) geometry. The curved surfaces send information out into the flow to tell it to adjust its flow direction. These are weak expansion waves causing the flow to increase in speed and drop in temperature. When this temperature reduction takes the air below the dew point the result is water condensation. The condensation region is not collocated with the shock wave, nor directly initiated by the wave.

A finer point to consider about the appearance of the cloud is that it takes time (milliseconds) for droplets to form so that the dewpoint line must be a bit ahead of the cloud. At the trailing edge, the recompression and the associated warming will cause the droplets to evaporate, again delayed by the finite rate of the process. The details are quite complicated by the additional consideration of the latent heat involved in the process, but they are beautiful to observe.

We note that a rocket ascending into space may experience the cloud formation phenomenon around its nose when the atmospheric conditions allow it. The physics is the same as described for our wing.

The above diversion into supersonic flow phenomena was necessary to explain the condensation observations shown in Fig. 7.9. We leave the subject, revisit it in Chap. 13, and return to the concluding discussion of subsonic airplanes and their performance.

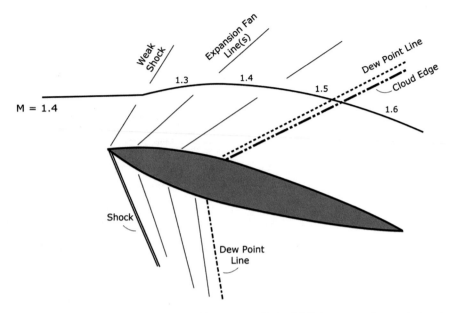

Fig. 7.11 The wave pattern associated with a more realistic airfoil (exaggerated geometry) and operating at supersonic Mach numbers. The values of M along the expansion waves are meant to be illustrative and depend on the airfoil geometry. The wave pattern under the wing is similar

7.9 All the Airplane Aerodynamics in One Place

The understanding we have built can be put in a form that is useful for description of the aerodynamic performance of a wing and/or an airplane. In addition to the lift curve described earlier, a good way of representing this summary is the so-called *drag polar* where lift is plotted versus drag. Figure 7.12 is such a plot for a commercial airliner. Lift and drag are both noted in non-dimensional coefficient forms using

Fig. 7.12 Sketch of lift and drag performance of a subsonic airliner type of airplane. This drag polar is typically obtained for various Mach numbers of interest. A genuine performance plot for a Boeing 727–100 is shown in Appendix D

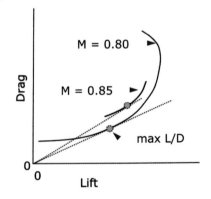

dynamic pressure and wing area. For an airplane designed to cruise efficiently, the drag per unit lift is very important because it determines the engine thrust required and the associated fuel usage. The drag polar for a commercial airliner is usually determined for a variety of flight Mach numbers. An aerodynamicist measures the efficiency of a flight vehicle or its wing by the *lift to drag ratio*, or *L/D*, this ratio is akin to the inverse of a friction coefficient determined in the high school physics experiment involving the dragging of a brick across the floor.

The tangent line from the origin of such a graph to the curve identifies the operating condition associated with maximum *L/D*. Piston-engine powered aircraft are typically operated at a speed for maximum *L/D* for good fuel economy which is especially important if long range is to be realized. For a modern commercial jet airplane, this ratio is in the vicinity of 20 meaning that the drag (and engine thrust) is about 5% of the lift. That parameter grew from about 15 in the early days of the jet age.

It turns out that a jet airliner operated at maximum *range* (to minimize fuel burn) has to operate at a point where the product of flight Mach number and *L/D* is a maximum. This is associated with the reality that a jet engine driven airplane engine produces thrust in contrast to a piston engine that produces power. A maximum value of *M·L/D* exists because increasing Mach number leads to falling airplane *L/D* due to the drag mechanisms associated with transonic flight as discussed above. As a consequence, commercial airliners typically operate at specific Mach numbers around 0.80. Because the jet airliner's best performance is sensitively connected to flight number (rather than speed as such), instrumentation for such airplanes includes a Mach meter.

Airplane design and operations were attempted in the past where cruise Mach number was meant to reach 0.90 for competitive advantage between airlines, but such performance goals did not meet with long-term success. For airplanes designed to operate at supersonic Mach numbers, the situation was even more dire. The modest values of *L/D* achievable was the Achilles heel for such airplanes. That was an issue with the now out-of-service Concorde and would likely have been a concern for the American SST (*SuperSonic Transport* and its contemplated derivatives) had it been built and put in service.

7.10 The Computer to the Rescue

Computers today play an enormous role in much of engineering, manufacturing, and commerce. Their role in understanding flow phenomena was and is central to build better models, to wit, mathematical models. The reality of the complexity of flow description and the motivation for overcoming it with nimble and automatic computing machines deserves a few comments.

The simple physical models that make it possible to visualize the important phenomena involved in air flow about airplanes are just that, simple. In our tale, references have been made to better analysis results from the output of computers. The reasons behind that warrant a little more exposure to daylight.

In the digital age, it is fair to say that just about all important determinations of the performance of wings and airplanes is done on computers. One could say that within the computer is a mathematical model of an airplane. So why the computer and what does it actually do? The computer is much less expensive to run than experimental tests using wind tunnels and certainly more accurate and faster than any mathematical modeling done with paper and pencil. The computer can quickly accommodate a configuration or operational change and improved to reflect a better understanding of the modeling architecture.

As far as flow analysis is concerned, the large modern computer is well suited to the task because the description regimen is based on a set of sometimes complicated differential equations and the field is large. The 'differential' aspect has to do with the mathematical modeling of the entirety of the flow field as an array of small elements, describable as cells, and writing Newton's equations of motion for that cell. The equations of motion are a form of $F = ma$ that relates pressure differences and shear stresses acting on a cell to the local change in velocity.

The geometry of an airplane or wing being examined is a boundary of the flow field, as often are the conditions very far away from the object. In internal flows, such as those through a nozzle or diffuser, the boundaries are hard surfaces external to the flow. Together, specification of such descriptive parameters forms what are called *boundary conditions*. Dominant among these is that the local flow velocity cannot be normal to a rigid surface, nor can it be anything but zero along the surface when friction effects are considered.

The basic calculational cells are small. Imagine, for example, the number of cells with a characteristic dimension like one centimeter or half an inch required to describe the flow about an airliner. It is huge. Further complicating the matter is that within flow features like boundary layers and wakes, the cell size there has to be even smaller. On the other hand, the cells far from the object can be larger. There changes in cell properties (velocity, etc.) vary gently, meaning slightly over larger distances.

Keeping track of properties in a large field requires large computers with high information storage capacity and rapid computation speed. Historically, the demands of computational capability for the analysis of airflow investigations motivated the development of heavy-duty computers. Another motivating and related application demanding such computers is the analysis of weather systems. The computer is well-suited to do the repetitive calculations required until adjacent air flow elements form the nice continuum nature requires.

The flow phenomena that influence wing performance such as viscosity effects and waves markedly complicate the descriptive equations so that dissimilar sets of equations govern the behavior of cells. These regions are the ones that may be heavily influenced by entropy generation mechanisms. In boundary layers, for example, the effects of friction involve the necessary inclusion of terms in Newton's equations of motion that account for friction. These unfortunately increase the difficulty of managing the mathematics, especially so when the flow is sufficiently energetic so that proper accounting of energy is critical. Luckily for the engineer, under many circumstances, the viscous terms can be safely omitted from descriptions in the far field. When the field consist of regions analyzed by locally different means, the computer analysis is often carried out in subregions. In the end, the results have to be stitched together to ensure a realistic variation of properties as required by nature.

As far as friction effects are concerned, an important challenge is the correct description of friction effects, be they laminar or unsteady and turbulent. Wave phenomena also present computational challenges because waves reach into the far field. Quite challenging is the situation where the effects of waves and friction merge into the need to properly describe the shock/boundary layer interactions that play such an important role in flow separation from external or internal flow boundaries.

A lot of progress has been made in computational fluid dynamics to allow the determination of good performance estimates. By good is meant, the calculation of integrated results like lift and drag, with narrow ranges of uncertainty under the many realistic conditions that might be encountered in practice.

7.11   Aerodynamic Heating is Not Due to Friction

Aircraft like the Concorde, the SR-71, or the Space Shuttle come in to land with an elevated skin temperature. The Concorde just spent 3 h cruising at $M = 2$, the SR-71 spent time at more than $M = 3$ and the re-entering Space Shuttle, who knows? Peak Mach number reached is said to be around 25 through all layers of the atmosphere. The skin temperatures reached by the two supersonic cruise aircraft (195 °F (90 °C) and 500 °F (260 °C), respectively) have a lot to say about the aircrafts' design. The skin and structure of the Concorde is made with aluminum and the SR-71 with titanium. The Space Shuttle was covered with ceramic tiles that can stand the hellish temperatures of re-entry from space and are replaced as necessary after landing. Are these material choices associated with the flight Mach numbers? Very definitely. Are they due to friction? The answer is No, with a little bit of Yes.

Let's look at the nature of friction. A friction force is experienced by a solid body (as our high school physics experiment again) whose motion is retarded by stationary surface like a brick sliding on the floor or a rope slipping through our hands. When a force, including a friction force, acts with a speed, there is an expenditure of mechanical power. In the case of friction forces, the mechanical power is converted to thermal power, heat. That is the nature of our experience with frictional heating and why automobile or aircraft wheel brakes get hot when used aggressively.

For fluids, liquids and gases, the situation is more complicated. Shearing forces acting on moving elements of fluid exert mechanical power that ends up in the fluid itself. For example, the oil in a journal bearing experiences such work expenditure and, as a result, the oil warms up. In many applications, such heating has to be dealt with by cooling of the liquid involved.

For unbounded fluids flowing past a surface, friction is manifest in the creation of a boundary layer on the stationary surface. Within the boundary layers, there are shear forces acting on adjacent moving elements of fluid. The heat generated is the product of this shear and the local velocity. At the bottom of the boundary layer, the shear is relatively large, but the velocities are small while the opposite is true near the edge of the boundary layer. There is, consequently, a zone of peak heat generation somewhere in the central region of the boundary layer. This heat is distributed by the motion of molecules and/or eddies within the boundary layer and plays a role (a very small one as it turns out) in heating the surfaces of our flight vehicles.

Is there another source of heat for the air passing along our high-speed airplane? Yes, and it is an important one. By its nature, the boundary layer air is forced to come to rest relative to the airplane, or at least slow down. The air that is so slowed brings with it its kinetic energy. When that kinetic energy source is large, as it is for high-speed flight, the local heating from the deceleration of the freestream air delivers kinetic energy converted to thermal energy that is not just significant, but dominant. That last statement has to be supported by plausible evidence.

Short of a detailed analysis, the best one can do is to make an order of magnitude assessment. That involves examining what physical quantities play roles and devising a figure of merit that characterizes the relative importance of two effects to be compared.

Such a figure is the ratio is of the heat generated by dynamic conversion of kinetic energy and the heat generated by the shear forces associated with friction. One can show that the ratio is closely related to and numerically similar to the Reynolds number (based on flow distance), a number that is numerically in the millions for a full-scale airplane. In short, frictional heating on a high-speed airplane is very small compared to the heating from the kinetic energy of the oncoming air. Engineers who have been concerned with this have established that the surface temperature on a surface like that of an airplane is typically somewhat less than the stagnation temperature of the air. Near a stagnation point (on the blunt nose) or line, however, like the leading edge of the wings, the temperature realized there is the full stagnation temperature. Thus, one can say, at least for conservative design purposes, that the entire vehicle is bathed with air close to the stagnation temperature.

This state of affairs is well illustrated qualitatively in the Space Shuttle sketch (Fig. 7.13) showing the leading edges white hot while the lower surfaces of the wings and body are also hot but cooler. NASA engineers equipped the leading edges as well as the bottom surface of the Shuttle with the best performance thermal protection system that they could devise.

Leaving the Space Shuttle aside, we can look at the temperatures experienced by the other two airplanes as representative examples and see how the stagnation temperatures at the conditions where they fly correlates with after-flight body temperatures. It is reported in practice that the Concorde operates skin temperature is typically between about 200 °F and 260 °F (93–126 °C) while that of the SR-71's is typically near 500 °F (260) and sometimes as high as 1000 °F (540 °C).

The stagnation temperatures relevant to these airplanes can be estimated by making a few assumptions. Consider that we are flying in an environment like the stratosphere where the absolute temperature is about 400° R (-60 °F or -51 °C). The Rankine scale is identical to the Fahrenheit scale except that it is referenced to absolute zero temperature. For practical reasons, namely that temperatures near absolute zero are not in our everyday world, the Fahrenheit scale, still used in the US, employs some diabolical thinking that the temperature of freezing water (at 14.7

Fig. 7.13 Sketch of the
Space Shuttle re-entering the
atmosphere (NASA)

psi absolute atmospheric pressure or 101,000 N/m^2) is exactly 32 °F. Because this
scale is familiar, at least in the United States, we will stick with it. Table 7.1 shows
the Fahrenheit temperatures reached at a stagnation point. What is apparent in this
table is that a Concorde flying at M near 2 will be hot on landing. More so for the
SR-71 that is capable of speeds greater than $M = 3$. Aluminum melts near 600 °F
(315 °C) while titanium melts at over 3000 °F (1650 °C). The Concorde design could
safely use aluminum, but the SR-71 had to employ a titanium structure. Note that
our commercial airliner's skin temperature is 20 °F (-7 °C), cold enough to face the
possibility of ice build up under some conditions. Their leading edges are equipped
with means to heat them and thus avoid worries with icing issues whose buildup
might change the shape of the airfoil shape that was developed with such care.

Now consider that the Space Shuttle reached Mach numbers near 25. It 'flew'
back into the atmosphere at speeds where surface temperatures reached are said to
be near 2300 °F (1260 °C) and air temperatures higher yet!

The numbers in the table are estimates. Nevertheless, the good correlation between
observed vehicle temperatures with the stagnation temperatures seems a sufficiently
strong to able to say, with a good deal of certainty, that friction per se plays a very
small role in contributing to airframe heating of these examples.

Table 7.1 Stagnation
temperatures for an airplane
flying in the stratosphere at
various speeds characterized
by Mach number

Aircraft	Mach	T(stagn, F)	observed skin Temp (F)
Balloon?	0	-60	–
Airliner	0.85	20	Not measured
Concorde	2	260	195–260
SR-71	3–3.3	660–810	500–1000

Chapter 8
The Jet: Fluid in Motion and More Vorticity

A jet is always involved in propelling any object through a fluid medium, even by living creatures. A propeller creates a jet and does it mechanically. The so-called jet engine does it by other means that are described later in our story. At this level of looking at the physics, there is no difference between piston engine driven propellers and a jet engine: both create jets of air.

The discussion of wings lifting an airplane noted that a vortex must form a closed loop or end on a boundary. A beautiful example of a boundary is a water surface under which we can create a jet with a paddle and observe the ends of a vortex loop (Fig. 8.1). Under the water is a "U" shaped vortex that is akin to the smoke ring blown by the person in Fig. 6.3. In the flight environment, such boundaries are sometimes harder to discern if they exist at all.

As we stand next to a person blowing the smoke ring (Fig. 6.3), the ring will be seen to proceed in the blown direction at a modest speed. That speed has to do with the strength of the vortex and its size. Specifically, the velocity induced by the vortex from the portion at 3 o'clock interacts with the portion at the 9 o'clock position and that interaction leads to force just like the bound vortex on the wing provides lift. Now imagine the person can blow many rings in rapid succession so that the vortices form a tube. We can describe a vortex tube as a jet boundary. Returning to the canoe paddle example, the jet can be visualized there if the paddler strokes the water in rapid succession. He will leave an aligned array of vortex pairs as evidence of the jet he created. This is precisely how a stern-wheeler ship is propelled. Taking the water example a step further, we note that a fish leaves alternating vortices at his left and right rear when it sweeps its tail from side to side. Such an array is called a *vortex street* forming a jet to the rear and one could say that the fish is jet propelled.

The vortex pattern from a closely coupled set of rings will resemble the pattern issuing from the tip of a propeller turning at high speed. The pattern from the propeller will be helical in shape and issuing from near the blade tip. With a large number of blades, the vortex approximates a cylindrical sheet produced by the propeller.

© The Author(s) 2022
R. Decher, *The Vortex and The Jet*,
https://doi.org/10.1007/978-981-16-8028-1_8

Fig. 8.1 Canoe paddle in water. Note the surface deflection associated with the locally higher rotational speeds near the center(s) (© JP Danko/Stocksy United)

8.1   Prop Tip Vortices

Let us look at how vortices are involved in the creation of a jet by a propeller. The propeller blade is like a wing that rotates around a hub. As in the case of a wing, there must be a bound vortex associated with the blade and a trailing vortex from the tip.

One might ask: 'where is the closed loop of the vortex on the hub side?' A vortex can end only on a boundary and the hub may just be such a boundary, albeit not as nicely defined as a wall or surface. A more realistic description of the flow near the hub is that the vorticity trailing from the propeller blade near its hub is shed just like that from the tip. Figure 8.2 shows the vortex pattern around a single propeller blade with its root that is exaggeratedly further out from the hub than a realistic blade would be. Because the trailing vortices are helical in shape, the induced velocities have axial (the jet to the rear) and azimuthal (in the plane of propeller disc) components. Thus, both hub and tip vortices contribute to the rotating part of the jet flow. The hub vortex is not readily observable as a bundled vortex like that from a wing tip or the propeller blade tip. The reason is that the turbulence allows for a ready sharing of the vorticity with the whole of the jet. In practice, the body of the airplane or nacelle of a tractor (as opposed to a pusher) propeller interferes with an organized survival of the vorticity shed at the blade hub.

The pattern shown in Fig. 8.2 would naturally be compounded by other blades of the propeller. While the vortex issuing near the hub would likely degenerate into less organized vorticity, the condensed moisture in the tip vortex of propeller can, under certain conditions, remain organized and visible as seen in Fig. 8.3.

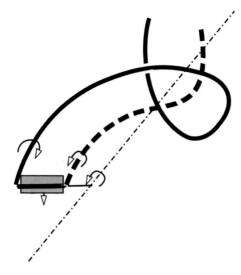

Fig. 8.2 The vortex pattern from an isolated propeller blade rotating about a centerline. The trailing vortices are helical due to the blade and air motion. The blade with an exaggeratedly large hub radius is moving toward the reader

Fig. 8.3 Two examples of tip vortices made visible by atmospheric conditions. At standstill the helix is tight and expands in forward flight as shown at right (Photo at right Belgian Air Force C-130H by Ron Kellenaers # 1131694, Airliners.net)

In contrast, the visible remainder of a turbofan jet engine in flight at altitude is a contrail. Its origin is, not from vorticity as such but, from the interaction of moisture laden engine exhaust mixing with the very cold atmospheric air at high altitudes. This is the case when there is a significant gap between the engine and the beginning of a contrail as shown, for example in Fig. 8.4.

Fig. 8.4 High altitude contrails from a A340 commercial jet airliner (Photo: Adrian Pingstone; curid = 1856927, http://commons.wikimedia.org/wiki/User:Arpingstone)

8.2   Jet Boundaries Are Cylindrical Sheets of Vorticity

Vorticity in a variety of forms plays a necessary role in propelling an airplane. From an overall performance viewpoint, the propeller's jet and that from a jet engine differ mainly in the amount of air processed and the speed of the jet.

In practice, the lift distribution arising from the rotating propeller blade will produce a somewhat non-uniform jet velocity because the lift distribution varies along the blade. However, the vorticity quickly assembles into a bundled vortex as the illustrations above show and the vortex sheet in the form of a tube eventually allows the jet internal velocity profile to be relatively uniform. The word 'sheet' may be described as the approximation of a sheet made of helical vortices from the ends of the blade. The jet from a jet engine tends to be uniform from the start because it exits a reservoir of pressurized gas by means of a nozzle with little rotation by virtue of the design of the turbine from which the propelling gas exits.

Unlike the wing that cruises into vertically still air and imparts downward momentum, the propulsion system must deal with the momentum associated with flight speed and direction. It must therefore manage the *energy* of the incoming airflow efficiently so that the installed heat engine will not have to make up for unnecessary losses. This requirement is more severe at higher flight speeds.

How does the propulsion system do that? In the case of a propeller, the blades deal with a combination of speeds: the rotational speed of the blade and the forward speed of the airplane. These two speed vectors determine the angle of incidence on the blade. Because rotational speed varies with location along the blade (faster further out), the incidence angle varies along the blade. The propeller blade is designed with twist to achieve a reasonably good distribution of the local angle of attack all along the blade. For airplanes that don't fly too fast, like modest cost general-aviation aircraft that may be limited to speed between 60 and 150 mph (27–67 m/s), a fixed twisted blade may be adequate. This is called a fixed pitch propeller (see, for example, Fig. 2.1).

For propeller airplanes that fly faster, a mechanism is included in the design of the blade that allows the blade as a whole to be rotated about its axis at the hub. This allows optimization of the blade orientation to be realized either by mechanical adjustment by the pilot or automatically by a mechanism that senses the appropriate speeds. The historical development of propellers took place from the inception of aviation to the 1940s with ever more complex pitch-control mechanisms for ever better performance.

8.3 An Aircraft Can Never Have 100% Propulsive Efficiency

… and, for the same reason, neither can a power boat! But your car can!

The key to success of the propeller as a thrust generation device for an airplane was that the speed increase in the 'jet' over the incoming freestream was low enough so that it was quite efficient. In other words, the kinetic energy investment by the propeller in the jet was modest. This notion underlies the idea that there is an efficiency that can be defined in connection with the creation of a jet by a flight vehicle propulsion system. The system is designed to create an increase in *momentum*, but the reality is that it produces power in the form of jet *kinetic energy*. Momentum depends on the velocity directly while kinetic energy depends on the square of velocity. That algebraic distinction allows stating the power delivered to the airplane relative to the power that has to be expended in order to realize the desired momentum increase. That fraction is called the propulsive efficiency of the system and is possibly the most important aspect of the system's utility because it is an important component of fuel usage. Without going through the algebraic steps, the propulsive efficiency is given by $2/(1 + V_j/V_0)$. The two symbols stand for the jet and flight velocities respectively. To express that in percent, multiply that fraction by 100!

8.3.1 A Quantitative Aside

The next two paragraphs describe the origin of this critical relationship for the interested reader. The laws of mechanics state that work is done by a force moving over a distance and power is work per unit time. Thus, the power exerted by an airplane propulsion system in steady level flight is the thrust times the flight speed. The thrust is the momentum *increase* exerted on the stream tube that enters the propulsion system. The initial momentum is air mass flow rate (m in lbs/s or kg/s) times the flight velocity. Thus, the thrust[1] is

$$m(V_j - V_0) \text{ and the power exerted is } m(V_j - V_0) \cdot V_0$$

The power provided by the propulsion system is the *net* mechanical power in the air of the jet:

$$^1/_2 m \left(V_j^2 - V_0^2 \right)$$

The ratio of these two power quantities gives the propulsive efficiency noted above.

We note here that the mass flow rate (m) is all the air processed by the propulsion system. For a fan jet, that includes both the fan flow *and* the primary or 'core' flow through the components of the engine proper. In that sense, the velocity of the jet is an average that may or may not be mixed in an actual turbofan engine.

One hundred percent efficiency is realized when the two velocities are the same. These two velocities are identical if you look at the wheel of your car being the propulsion system and the almost infinite Earth is the medium. That also means that the momentum *increase* of the medium is close to zero meaning that by moving our car, no energy is left in the medium. This silly example is an accurate description of a wheel (a propulsion system) in contact with the ground where we enjoy one hundred percent propulsive efficiency, provided we don't spin wheels and thereby shoot gravel at a neighbor's window.

Typical propulsive efficiencies for propeller driven airplanes range around eighty percent and better, depending on the sophistication of the geometric design and operation of the propeller. Arguments were made against the practicality of the jet engine because the exhaust velocity could be calculated and determined to be quite high. For a nominal and realistic (at the time) speed ratio (jet to flight) of 3 to 1, the efficiency is just 50%, judged to lead to an excessive fuel consumption rate by the engine. That argument was ultimately judged to be false (or at least could be overcome) and proponents did manage to build what we call a jet engine.

8.4 Counter-Rotating Props

In addition to this aspect of efficiency, the conventional propeller imparts a rotation to the air processed by it by virtue of the drag forces similar to those on a wing: pressure, viscous and induced drag components. We have suggested in the earlier text that the induced drag is responsible for the rotation of a propeller's jet. Such rotation represents rotational kinetic energy in the 'jet' stream behind a propeller. If nothing else is done, this energy is wasted and presents an additional loss that,

---

[1] For an accurate statement regarding engine thrust, there may be a pressure term in addition to the momentum term. In air-breathing propulsion this term is usually small and neglected here. Not so in connection with rocket engines discussed in Chap. 13.

Fig. 8.5 The British Bristol Brabazon airliner prototype and its piston engine installation for counter-rotating propellers (Photos courtesy BAE SYSTEMS)

if sufficient in magnitude, might have to be dealt with. It contributes nothing to the propeller thrust and is eventually dissipated in the turbulence of the jet.

In a number of aircraft with propeller engine configurations, the loading of the propeller was sufficiently high that something had to be done to recover the energy in the rotation of the 'jet' stream. In the waning days of piston engine technology for large commercial or military airplanes, the possibility and desire arose for ever more powerful engines to drive propellers. These promised faster and larger airplanes, but they also pointed the designers to the need for fuel use efficiency to achieve long range. The wasted rotational kinetic energy had to be recovered.

In order to put more power into the 'jet' and realizing reasonable fuel usage while keeping propeller diameters reasonable, airplane builders turned to counter-rotating propellers. A second set of blades is tailored to take the rotating output from the first and add power in such a way as to leave little residual kinetic energy in the jet's rotation motion. The nature of the jet flow behind a propeller is not nice and clean, so that to placing two propeller sets in tandem in this way is challenging to do well, but it is somewhat effective. Examples of such configurations are the British Bristol Brabazon (Fig. 8.5) and the Tupolev Tu-95 strategic bomber built in the Soviet Union during the Cold War shown Fig. 8.6. The Brabazon airliner used two piston engines[2] to drive each propeller set while the Tu-95 used a single gas turbine engine. In both cases, four set of two propellers each were used, two on each wing, eight engines in total. Gearing was necessarily required. The concept never went very far for economic reasons and gearing always comes with weight, power consumption, and reliability issues of its own. Further, an airliner conversion of the Tupolev bomber proved the concept to be very noisy for passengers. The turbojet age was dawning and was about to change history. In the US, the counter-rotating propeller concept was never brought to a widespread reality. One place where it was considered was for a number of vertical takeoff military fighter prototypes (Fig. 8.7). For such airplanes

---

[2] The eight power plants for this airplane were Bristol Centaurus engines that sported sleeve-valves.

Fig. 8.6 The Cold War Soviet Tupolev Tu-95 strategic bomber using counter-rotating propellers

Fig. 8.7 Convair XFY-1
"Pogo" prototype, a vertical
takeoff and landing fighter
(1954) with counter-rotating
propellers powered by a
turboprop engine. A similar
Lockheed airplane was built
for a competition that did not
lead to a production contract
(US Navy)

to leave residual rotating flow in the lifting jet means that the airplane will undergo a rotation as a reaction. This may present an undesirable control issue.

The idea of counter-rotating blades is not singular in propeller-like applications. The gas turbine will be explored in detail further on, but this is a good place to mention that the idea of counter-rotation arose in connection with gas turbine engine and its compressor. For example, during the earliest days of gas turbine engine developments in Germany (late 1930s), the thought was explored that, for a lighter weight and shorter compressor, it could have a counter-rotating set of blades rather than a rotor-stator combination. It was deemed to be too complex and abandoned. Another stab at this idea was undertaken in the US in the 1970s when government contracts were let to explore the feasibility of a cowl-less turbofan for commercial airliners. This engine type resembled a counter-rotating propeller and was built around the gas turbine engines of the day. Major manufacturers tried the idea. They were called 'unducted fans' or 'propfans' and were considered for proposed airliners. These would have been constrained to flight Mach numbers around 0.7 because of the fan tip speed limitation. They found no application and the effort was deemed to be a dead-end, at the time.

The reader might wish to examine archival information concerning an attempt to operate a propeller at supersonic speeds. This is possible given sufficient power. Such a propeller was incorporated in a military airplane designated as the XF-84F. It turned out to be a spectacular failure and is said to have been the loudest airplane that ever flew!

Chapter 9
Propulsion for Flight: Power or Thrust?

The modern jet engine is a bit mysterious. Does it work like the engines in our cars? How and why is it better than the piston engines of the time during and after World War II? The obvious answer has to do with the ability of modern airplanes using jet engines to fly faster, higher, and farther. The reasons are interesting from historical and technical viewpoints and our focus will center on the latter. That will involve extending the ideas developed for understanding the function of the wing. To wit, the energy conservation law developed for the behavior of the flow along a streamline will have to be used in a more general form to include manipulation of the air when we compress it or allow it to expand to do something useful. After all, we want the engine to be practical and get work out of it. The heretofore assumption of adiabatic flow, i.e., flow without heat input, will have also to be revisited for the combustor because we are going to add fuel to air and burn it!

As the wing produced momentum as downwash that we exploited for lift, the propulsion system's function is to produce momentum of its own. The reason is identical: a force results from changes of momentum and we will want to do a lot of that in order to be able to propel an airplane forward at high speed.

So, what is a propulsion system? It is nothing more than the means to take in air and expel it at higher speed in the form of a jet. In that sense, it is a machine that pumps air. The propeller-driven airplane carries out that function by rotating wing-like elements with power input from an engine that produces mechanical power. The air used for propulsion does not participate in the production of the power. By stark contrast, the jet engine intimately intertwines the two functions of power production and propulsion. The jet engine does it in mysterious ways that involve the laws of heat physics raised earlier in connection with flow along a streamline. The pros call that thermodynamics, but don't let that term be scary!

For brevity, I will refer to the Internal Combustion Engine or equivalently, the Otto cycle piston/cylinder gasoline engine, as an ICE. For simplicity, the discussion of the jet engine functionality will focus on the basic turbojet. Thus, the terms 'jet engine' and gas turbine engine are, for our propulsion purposes, equivalent. Extension to consideration of turbofans and other gas turbine engine types will be made further on.

© The Author(s) 2022
R. Decher, *The Vortex and The Jet*,
https://doi.org/10.1007/978-981-16-8028-1_9

First let's tackle the two engines and see how they are different and how they are similar. Both are *heat engines* that produce power from the heat obtained by burning fuels. The power is ultimately manifest in the propulsion jet produced. In both cases, the goal is to produce the thrust necessary to overcome drag. A distinction between the two engines is that the power output from the ICE is *shaft* power whereas the power in the jet engine resides in the gas jet produced. The distinction between interest in power and thrust may be ascribed to a historical evolution of book-keeping associated with the reality that developing the piston engines was logically treated as separate from developing propellers. For the jet engine, the power produced is less interesting than the thrust, hence a lesser interest on power per se. That is not to say that it is unimportant because it is the centerpiece of any discussion of *how well* the jet engine works to produce thrust. This dimension was tackled in connection with the discussion of propulsion efficiency.

9.1 The Engines: Stark Differences

The piston engine (ICE) processes air in piston/cylinders where the air processed is taken in batches through a series of *time separated* events. By contrast, the gas turbine engine carries out similar, though not identical, events in a *spatially separated* set of components that operate separately, steadily, and continuously. That is the first important difference between these two engines. The second is, as stated above, that the air processed by the ICE is generally not part of the propulsive jet whereas the air processed by the gas turbine is the jet, all of it.

The ICE engine therefore operates independently of the flight speed of the airplane in which it is installed. It must, however, deal with the static atmospheric environment of the airplane, namely the density of the air at the altitude where the airplane is flying. Consequently, the ICE performance depends primarily on altitude. At higher altitude the atmosphere is thinner (less dense) than at sea level. At these higher altitudes, the finite volume of the engine's cylinders takes in a correspondingly reduced amount of air mass with each intake stroke of the piston/cylinder. The necessary reduction of fuel to burn with the diminished amount of air reduces power and that, in a nutshell, is the Achilles heel of the piston engine. The ICE's development was nevertheless essential for practical flight because, for the first time, an engine light enough and powerful enough could do the job. Much effort has been spent on mitigating the altitude limitation with superchargers and turbochargers that have been effective, but never to the extent one might wish.

The turbojet engine is quite different. All the air used for propulsion is processed by the engine. That fact places a premium on efficient air handling in every component of the engine. The net distinction between the two propulsion systems is that the ICE/propeller jet's large diameter and modest ability to add momentum to the air contrasts starkly with the jet engine's jet high velocity and rather small volume flow rate (cubic feet or cubic meters per second).

What kind of thinking went into the idea of a turbojet? The history reveals similar thoughts by engineers in Germany and Great Britain during the 1930s and 40s. Names associated with this engine development include the principal contributors Frank Whittle in Great Britain and Hans von Ohain and Anselm Franz in Germany. Others, of course, helped lay the foundation for what these men brought to reality. This history is well told elsewhere and our focus here is on the technical underpinnings related to the function of this engine.

9.2 Limits of the Old

To begin, we explore the workings of the piston engine. The ICE produced mechanical power by compressing air in a piston-cylinder, burning fuel in the compressed space, and realizing a rapid and great increase in pressure. The piston then allowed expansion with a much larger force on the piston and therefore larger power output from the piston; importantly, more power than it took to compress the air. The process of compression and expansion in the confines of a piston and cylinder (hence the word *internal*) was, at the time, the best way of handling the gases involved. In the ICE, combustion takes place in a very short time while the piston hardly moves. This combustion process is termed to be at *constant volume*[1] and the goal is to raise the gas pressure.

The efficient compression process by means of piston within a cylinder took time to develop. As rings, seals, valves, etc., were successfully improved, they were quickly adapted to propelling automobiles and, in short order, aircraft. The ICE rapidly supplanted and displaced the only other way to produce mechanical power, namely the steam engine, an *external* combustion engine because the heat is supplied to the pressurized water from the outside. Its virtue was the ability and ease to compress (or perhaps better, pressurize) liquid water. This process was and is easy to do with a pump and allows the generation of high-pressure steam to act on a piston. The pressurization of the water did not require large amounts of compression power. That was pivotal in the steam engine becoming the first practical way of generating mechanical power from heat.

The ICE does not offer many options for improving performance. The performance measures are usually the power output, the fuel consumption rate, and the engine weight involved. The compression is limited by the volatility of the fuel in the fuel–air mixture being compressed lest it ignite before the piston reaches the so-called top dead center and result in potentially damaging pre-ignition. The main challenge for this engine type is to maximize the amount of air processed efficiently to realize a large power output. This feature is central in determining power output. Another challenging aspect of the design is the valving to manage flow in and out of the cylinder without overheating the valves themselves. After all, the exhaust gas

---

[1] The piston moves a little during the combustion, however, this descriptive notion is a good approximation of reality.

from the ICE is hotter than practical materials can easily handle in a continuous exposure. A number of valve configurations were tried and used, chief among is the common poppet valve and the less common sleeve valve pioneered by Bristol in Great Britain. The flushing of the cylinder space by fresh, cool air after combustion mediates internal surface temperatures and effective cooling of the external structure of the cylinder are key features of any design that allows it to be successful.

The practical option available for increased power is increased volume (the displacement) swept by the piston(s). That makes the engines larger and heavier. Nevertheless, progress in increasing power output per unit engine weight improved steadily over the years that these engines were the power source for all airplanes in aviation's youth. The zenith of their time is characterized by their production of one horsepower per pound of engine weight or one horsepower per cubic inch of piston displacement. One of the important means of achieving such performance levels was supercharging that primarily helped overcome the limitations imposed by the low density of air at high altitudes. In supercharged engines, the power necessary to drive it was obtained from tapping into the power from the crankshaft for a net gain in power from the engine as a whole. There was another way of getting such performance improvements and these were directly linked to ideas about the jet engine.

The ICE produces a high-pressure gas exhaust because the very high-pressure, very hot gas can only be expanded so much by a piston designed to compress the incoming air with the same piston stroke. The noise and hot, high-pressure gas emanating from these engines are evident manifestations that there is a lot of energy in the exhaust that might be used productively. Indeed, attempts at increasing engine power by taking advantage of the situation were developed: notably turbocharging (with or without intercooling), and the related turbo-compounding where exhaust turbine power is added to the power output shaft by appropriate gears.

Figure 9.1 illustrates several aspects of a supercharged piston engine for aircraft. The valves are the site where a limit on air processing capability is imposed on any piston ICE engine by limiting its rotational speed. This limit is a consequence of the relatively small flow area that can be made available by an open valve and the flow choking[2] in the presented passage. In the image, the green duct carries the incoming air to a poppet valve at right. The exhaust valve is at left. The lower picture is of the same display showing the radial flow supercharger (compressor) impeller with its green outlet duct funneling air to the intake valve. The impeller operates at many times the crankshaft rotational speed, hence the necessary gearing to drive it. At right are two of the rearmost set of pistons with the air cooling fins, cylinders, and crankshaft connections. The cam races for actuating the valves are shown but hard to discern. This Pratt & Whitney R-4360 engine is built with four sets of seven cylinders, 28 in all, and produced about 4000 hp. Parenthetically we note that the cooling air takes about 10% of the fuel's heating value away as a total loss that cannot be utilized for power production.

---

[2] See comments on choking of flow through nozzles in Chap. 13.

Fig. 9.1 The complexity of the piston aircraft engine and a dream world for mechanical engineers; here sectioned views of a P&W R-4360 (Pictures from an exhibit at the Udvar-Hazy Center of the Smithsonian National Air and Space Museum)

Naturally, there were improvements made throughout the history of its development, but the fundamental limitations were inherent in the engine. Breakthrough improvements were not to come. There was a new engine on the horizon.

9.2.1 The First Jet?

Before leaving the ICE, it is interesting to note that building and flying a "jet" airplane with an ICE as a power source for a compressor was considered by the innovative Italian airplane-building count Gianni Caproni. He built an airplane (the Caproni Campini N.1) using a piston engine driving a compressor that led air to an afterburner and a nozzle. He believed this airplane to have been the first jet powered airplane. It flew on August 27, 1940, exactly one year *after* Hans von Ohain's airplane, the first gas turbine powered jet airplane that flew on August 27, 1939. That flight was kept a secret by the German government and the Heinkel company that built it.

So, why use a new jet engine? The answer is simple: we want to fly faster! While Caproni's attempt to do away with the propeller was noble, it failed. It was, however, the right thing to try because it is the propeller that imposes a speed limit on the airplane. The flight speed of a propeller powered airplane cannot come close to approaching the speed of sound. The power supplied by an ICE are simply inadequate for an approach like that of Caproni. This reality underlies people thinking about "jet" engines to fly faster.

The road to the jet engines in use today is mile-stoned by a succession of inventions, innovations, advances in the understanding of the physics, and the reality of everyday operations. These innovations are well covered in the literature and we leap directly to the turbojet of the 1940s and its successor, the turbofan twenty years later. What is somewhat skipped over in this discussion is the early use of the radial flow compressor that allowed jet propelled flight to be realized but rather quickly faded into a historical footnote. This now uncommon compressor type did break the ice and pointed the way to a new engine type.

9.3 The Gas Turbine or 'Jet' Engine

Describing the jet engine is a bit like describing the human body. Where do you start? The heart, the brain, the skeletal structure, its function, or what? The jet engine is, mercifully, quite a bit simpler but not straightforward. It is also a *system* that has components that have to function together for the engine to perform well. Before we can talk about the components, we have to develop tools that allow a somewhat quantitative discussion of what happens to the air as it streams through a gas turbine engine.

When one contrasts the ICE and the gas turbine as engines, it is apparent that the ICE relies on constant volume combustion carried out in a few thousandths of a second, with compression, combustion and expansion carried out in the same space. The alternative way for burning fuel to generate heat and, with it, run an engine, is to employ steady combustion from a continuous source of heat. Thus, separate components are called for. The idea behind the gas turbine engine is to process large amounts of air so that large amounts of fuel can be burned for the realization of large amounts of power. The compressor to furnish this compressed air could have the absurd form of a piston-cylinder machine as used in industrial processes, but that would entail the air flow rate limitations inherent in the ICE and be unacceptably heavy. Could a compressor based on the idea of a rotating wheel accelerating air to higher speed and then slowing that air to realize high pressure be made to be useful? The answer is yes, because the dynamic pressure of flow is raised behind a propeller by virtue of the work input into the propeller. But how to do that effectively? How to raise the pressure by a much more substantial amount? And how to provide the necessary power?

One of the ideas that had to be part of the thinking about this new engine is that a turbine can produce many times the amount of power that an ICE can. The turbine is a steady flow machine with which there was a lot of industrial experience in hand. Perhaps a steady flow compressor and a turbine can be made to work together.

9.4   Is the Gas Turbine Engine Like Another Familiar Engine?

Suppose for a moment you are standing in front of a jet (or equivalently, a gas turbine) engine sectioned to expose its inner workings. The natural question "How does it work?" is often answered in terms of the similar processes in an automotive ICE. That may not be the best way of doing so, particularly because the heat addition part of the engine, as embodied in the combustion chamber, is nothing like the related process in an ICE.

The fundamental processes of compression, heating, expansion, and heat rejection must be present for all types of heat engines. It is the way that heat is added to the engine that differentiates the ICE and the gas turbine engine at a fundamental level. The work processes of compression and expansion, be they by piston or aerodynamic means, are a secondary but important *mechanical* distinction between these and other engines in common use.

The ICE is unique in that it operates with the fundamental processes listed above in a time-separated fashion in the same space. The gas turbine and the steam engine operate with these processes with separate components and are much better for comparison purposes. These two engine types operate with *steady* flow of their respective working fluids, air and water (as steam). To first order, a thermodynamic analysis of any engine does not care whether the processes are steady or sequential, just as long as they are present.

9.4.1 Yes, the Steam Engine!

Let us look at what goes on in a steam engine so we can relate events there to those in a jet engine. For our purposes, the steam engine model to have in mind is of a steam railroad locomotive that might commonly have plied the rails in the years between 1850 and 1950. It is very much a forebearer of the gas turbine engine even as the medium processed differs substantially from the air we might wish to use in a flight propulsion engine.

While most discussions of the thermodynamics of any heat engine invariably start with the compression part of the cycle, we will start with the biggest and most important part, the heat addition aspects. In the steam engine, the heat addition to the pressurized water is steady as the water 'flows' by the heat source. Here heating occurs in a large flow-through vessel, a boiler and a steam superheater. The boiler is fed by high pressure water and serves to create a very much larger volume of steam at the same high pressure. In the boiler, each cupful of water becomes about a hundred of such cupfuls of steam!

The resulting steam provides power by letting it expand against a piston within a cylinder connected to the wheels of the locomotive. That is how it generates mechanical power.[3] The power can also be extracted by expanding the steam through a turbine. The machinery for doing so has been employed for the production of electric power since the 1920s. The word "expanding" is used here because, in either work extraction process, be it in a piston/cylinder or through a turbine, the volume of the steam increases to allow work to be done the piston head or the turbine blades.

In order to make the argument that the work available by a greater volume of air plausible, consider the following thought experiment that leans on the use of pistons and cylinders. A piston within a cylinder made to undergo a certain amount of compression displacement, i.e., over a fixed stroke, requires work. Such a compression process can be very close to reversible because it can be designed so that the gas temperature is increased by compression does not lose any of the heat. In practice, good thermal insulation or a rapid movement will allow the process to be approximately adiabatic and reversible. If this piston is subsequently allowed to expand back to the starting point, then the work invested is recovered and, as far as the universe is concerned, nothing happened.

Now let us pretend that we have invented a piston and cylinder whose diameter can be made larger at will. This is improbable but not impossible. If the air in the compressed space is heated by some means (including combustion of fuel) in such manner to keep the pressure constant by letting our magic piston/cylinder grow as heat is added, we end up with a larger piston which is acted on by a larger force (because the piston area is larger) than the force required to compress the gas before heating. Allowing that piston to expand through the same stroke as executed in compression leads to a greater work output than was required by compression. The result is a net production of work for an input of heat. This is the way to think about an engine

---

[3] While the steam is admitted periodically into the cylinder, the process could be regarded as steady over a longer observation period.

with constant pressure heat addition, … if one insists on dealing with pistons and cylinders. At the end of the expansion process, the air is hotter than it was before compression and the heat associated with that is waste heat, a necessary consequence of the production of work from heat by any engine. This argument is, of course, an invocation of somewhat realistic magic, so let us return to real engines.

Before going on to the gas turbine, we note that the pump required by the steam engine is small in its role because water is incompressible so that raising its pressure does not require much in the way of power or machinery. That fact alone made the steam engine relatively easy (mechanically and in terms of the power required) to devise and allowed it to be the first heat engine of the industrial age. The supply of high-pressure air to the combustion chamber of a jet engine will not be nearly as easy.

In a gas turbine, heat is added to air by injection of a small amount of fuel into the compressed air stream. For all practical purposes, the combustion process that takes place within the engine just creates hot air. After all, air is mostly chemically inert nitrogen. The heat provided in a gas turbine combustion chamber typically increases the gas volume (per unit mass) by a factor of 2–3 times (depending on design conditions) the volume of air entering the combustion chamber. This much lower volume expansion sets the gas turbine engine apart from a steam engine. Further, in the gas turbine, there is no phase change from liquid water to steam. The *gas* turbine engine processes only gas (and tiny bit of liquid fuel).

Is the steam engine able to provide power for flight? Not likely. The steam engine requires lots of water and water is heavy as is the boiler. In a locomotive application, resupply of water emitted as spent steam is as important as 'refueling.' The locomotive carries tons of fuel and water in a tender right behind the locomotive. The weight considerations associated with the use of water preclude the steam engine for aviation. There are other issues. Water has properties that limit its performance in a steam engine. Most important of these is that the amount of heat necessary for making steam is very large. That has an important negative impact of the efficiency of the engine as a power provider. These realities did not deter potential aeronauts in the mid-nineteenth century from thinking about powering flight vehicles with the steam engine. It was the only engine available then. Such dreams were, however, neither practical nor realized.

The attraction of the gas turbine is that it works with *air* that is readily available for a flight propulsion engine and does not have to be carried along. That suggests a great advantage that an engine using air could be light in weight. Indeed, it worked out that way.

In order to examine the jet engine in greater detail, we have to step back and look at the way we have to describe what happens as air proceeds through its components. Thus, a brief pause to re-examine the parameters and the principles we developed in connection with flow along streamlines.

9.5 Energy Conservation Again, for Steady Flow Through Engines

For wings and external flows, we considered situations where the geometry deter-mines the condition changes along a streamline. The energy conservation statement can also be applied to *internal* flows where the flow is bounded by walls and perhaps manipulated by work or heat interactions. Instead of describing flow along a stream-line, the words and ideas will change to apply to a *stream tube* which is really nothing more than a bunch of streamlines undergoing the same or similar changes. The walls will play roles similar that played by a wing surface in that flow area changes will be called for to slow or speed the flow. To generalize the First Law, we extend it to apply it to situations where mechanical power (or work per unit time) is added to or removed from a stream tube. This is what is done to flow through a compressor or a turbine. Another extension is required to include the possibility of adding or removing heat from the flow. That extension is necessary for the description of flow through combustors or heat exchangers. The properties that reflect such interactions are total enthalpy (or equivalently, total temperature) and, depending on how well it is done, the total pressure. These interactions will be central to the functioning of a jet engine.

9.5.1 A Little Mathematics, Briefly

For the technically oriented reader, the next few short paragraphs are meant to display what the First Law looks like (as an equation) for the more general (steady flow) situations and how it applies to components of a jet engine. Inclusion of the mechanical power and heat terms transforms the First Law to reflect total temperature increases in direct measure of work addition and/or heat addition to a component. The inlet flow condition is described as state 1 and outlet as state 2. The engineer would write this for a unit of mass as:

$$C_p \cdot T_{t2} = C_p \cdot T_{t1} + w + q$$

where $C_p \cdot T_t$ is the total enthalpy (per unit mass, C_p is the specific heat), w is the work, and q is the heat input. Normally, both work and heat inputs would not be involved in the same component. The work term (w) would apply to compressors and turbines while heating (q) applies to combustors (or heat exchangers).

Depending on how efficient the work processes are carried out in the relevant work components, the total pressure change may be as good as dictated by an isentropic process. No real component is ever that good, but the technology is reasonably close to achieving such levels of performance. Thus, the total pressure increases in a compressor a bit less than ideally and decreases in a turbine, a bit more than ideally.

In a combustor or heat exchanger, the transfer of heat generally leads to a reduction in total pressure and good design aims to minimize it because the work components have to make up for such losses. For descriptive purposes, it is generally acceptable to talk about heat addition in combustors as taking place at constant pressure.

So much for the technical underpinnings.

9.5.2   What Happens in a Real Jet Engine?

In order to set the stage for the thinking about the new engine type, we look at some numbers related to the first jet engine in service, the German Jumo 004 flown in the Messerschmitt 262 jet fighter during WW II. The rather modest compressor (pressure ratio 3.1) required about 4000 hp from the turbine component for the engine as a whole to provide 2000 lbs of thrust. The turbine power output was substantial; it was certainly greater than what a typical piston aircraft engine could deliver. Thus, the turbine becomes an important component of the engine. With such thinking, the basic configuration of the gas turbine engine took shape.

The idea was that the compressor would pressurize the gas to a pressure *higher* than required for the jet so that the turbine can power the compressor and there would sufficient pressure left over to produce a jet.

The sketch in Fig. 9.2 illustrates the variation of the important (total) pressures and temperatures through a gas turbine engine. These values would be measured by a probe that brings the flow to rest locally. As a result of mechanical power addition, these flow parameters increase in the compressor (C) and decrease in the turbine (T). In the combustion chamber, or burner (B), the gas temperature rises and the pressure stays close to the value at entry. This interesting dimension of the combustion process is examined in greater detail in Chap. 12.

The temperatures and pressures reached in the engine are critical indications of engine performance. The pilot of a jet engine powered airplane is informed of engine performance through the instrumentation in the cockpit. The relative (total) pressure relation (turbine outlet to compressor inlet, see figure) is displayed as Engine Pressure Ratio (EPR), necessarily greater than 1.0 when the engine is operating. This ratio is the most important indication of engine performance as a thrust producer.

The mechanical integrity of the engine's hot section is ideally made by measurement of the turbine inlet temperature. Because of the high temperature at this point in the engine and the probe location, the measurement is difficult to make reliably in the long run. A measurement at the turbine exit (or in mid-turbine) is easier. It can and does serve as an indication of the temperatures in the hot section of the engine. The

Fig. 9.2 Simplified schematic variation of total pressure and total temperature in the principal components of a gas turbine or turbojet engine. The dots are the quantities involved in the cockpit display of engine performance

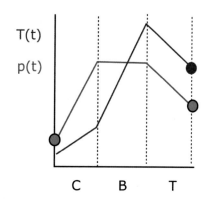

gauge display for the pilot is usually labelled as Exhaust Gas Temperature (EGT) or
Turbine Exit (or Outlet) Temperature (TET or TOT).

9.6 Design Issues

This idea of an engine arrangement was the focus of the pioneers in the late 1930s.
It required a number of difficulties to be overcome at the same time. The basic ideas
about what is necessary were available, but could they be combined into a practical
engine that can outperform the ICE? The first issue that had to be tackled included
devising a way compressing lots of air, efficiently. As it turned out, that proved to be
the most challenging of the issues facing a successful engine design.

A second issue was adding heat to the compressed air by mixing it with fuel and
burning it. At the time, industrial practice involved combustor chamber volumes that
would be unacceptably large for an airplane engine. The combustor volume would
have to be reduced.

A third issue was, and still is, that the temperatures reached by burning hydrocarbon fuel with air could be very high, higher than what metallic materials can
stand for prolonged exposure. Here the challenge is the maintenance of structural
integrity of the turbine by virtue of the temperature and exposure to the combustion
gas involved. The metal surfaces will have to tolerate the hot combustion gas and
last for acceptably long lifetimes. The turbine was, nevertheless, seen as practical
because the technology resembled that used in the steam turbines operating at that
time. Their aerodynamic performance was pretty well understood but the nature of
combustion gas as the medium was a new challenge.

So, the stage was set. The industrial world had experience in a number of related
areas that might be exploited for the process of building this new engine. Specifically,
experience with turbocharging of aircraft ICEs was in hand. In these devices, ICE
exhaust is made to drive a charging compressor to allow the engine to operate with
greater air mass than with normally aspirated engines. In fact, the air path through a
new 'jet' engine resembles that through a turbocharged ICE with the difference that
the heat source *is* the ICE. In short, the turbo-compressor feeds air into something
that heats it and the resulting hot combustion exhaust gas is expanded in a turbine to
drive the charger. The turbine exhaust is normally not used for jet thrust. The power
in that stream is too low to be useful.

In our telling of the story of the evolution of the jet engine, we will take the
viewpoint of emphasizing the engineering ideas and connect them to some of the
people involved. The people involved, Frank Whittle in Britain and the engineers in
Germany, were on opposite sides of a competition and later a war, so that information
sharing between them was very unlikely. Their respective governments classified
work related to the development as state secrets. Nevertheless, engineers were guided
by the physical realities and tended to think along parallel lines. Consequently, we
can develop the thinking about the gas turbine engine while jumping back and forth
across the war divide and thereby see how the necessary understanding grew into

practical engines on both sides of that chasm. While this discussion may feel like a departure from one concerned with vortices, the loop will be closed with ideas about the jet engine that have important operational similarities shared with the behavior of wings.

In the next chapter, we examine the compressor. It was *the* most important component of the engine because the compression process is difficult.

Chapter 10
The Compressor: Gas Turbine Engine Keystone

Compression of air by dynamic means was a known technology. The supercharger impeller in the R-4360 shown in Fig. 9.1 is a radial flow compressor that can achieve pressure rise factor on the order of three or so, a level that is sufficient to make a jet engine practical. For such an engine, the compressor impeller wheel just had to be made sufficiently large in size. It functions by taking in air and hurling it out in a radial direction with speeds that approaches the sound speed. The outflow air is then gently slowed so that the air's kinetic energy is converted to pressure. A flow tube with a gradually increasing area (called a *diffuser*) does that with an adverse pressure gradient as discussed in connection with flow separation from the upper surface of the wing. Since the diffuser has walls, boundary layers grow on them and they are subject to the same flow separation issue as we had on the upper surface of a wing.

The impeller of the radial flow compressor has problematic aspects that include a rather large surface area on the impeller and its guiding vanes. Thus, friction is experienced on those internal surfaces with an impact on efficiency. Another is the awkward geometry where the diameter of the compressor as a whole has to be about 3–4 times the size of the inlet duct. That limits its application into a sleek airplane or engine nacelle. Finally, the mass flow rate is rather modest but, in practice, that was partially addressed by having the impeller have two sides into which air flows. On the positive side, the compressor worked quite well. Frank Whittle and Hans von Ohain both used this kind of compressor on their early engines. Figure 10.1 shows a General Electric an I-16 and a later I-40 engine derived from the cooperation between Great Britain and the US during the war. Both engines have double-sided radial flow compressors.

The I-16 engine had a military designation of J31, with an airflow rate of 33 lbs/s (15 kg/s), a compressor pressure ratio of 3.8, a turbine inlet temperature of 1220 °F (660 °C), and produced 1600 lbs[1] of thrust. This was a small engine. The turbine shaft carried about 3400 hp to the compressor. The power in the jet at sea-level static conditions is about 2400 hp. These numbers are quite similar to those of the Jumo

---

[1] 1000 lbs of force in the metric system is 4.45 kN.

© The Author(s) 2022
R. Decher, *The Vortex and The Jet*,
https://doi.org/10.1007/978-981-16-8028-1_10

Fig. 10.1 Top: A General Electric I-16 (military designation J31, ca. 1942) engine on display at the Smithsonian Air and Space Museum. Note the double-sided radial flow compressor. Most of the diffuser ducting has been cut away but the air passages can be discerned. They include a 90° turn into the combustor. A better view of the diffuser ducting can be seen on an earlier Whittle engine in Fig. 12.5. Further diffusion takes place at the combustor inlet with a large flow area increase. A single stage turbine driving the compressor exits flow to the left. The lower figure is of a more advanced (early 1943) jet engine of a similar design (jet flow to the right) but with an in-line rather than reverse flow combustor (GE I-40, also known as the (Allison produced) J33) (Courtesy General Electric)

004 engine cited earlier. To put these numbers into perspective and appreciate what a technology step that the gas turbine was, consider that the P&W R-4360 engine, the largest production ICE engine that ever saw service application, produced an output of over 4000 hp processing about 9 lbs/s (4 kg/s) of air (at sea level). These engines are very different even with the gas turbine in its infancy.

We could discuss the radial flow compressor type in greater detail, but the reality was that another type proved to be superior. Part of the reason is that an analysis of the engine as a propulsion device points in the direction of a need for higher pressure capability from the compressor. A multiplicity of radial impellers operated in a series would be possible but quite awkward in practice for it would mean processing the air radially in and out a number of times. That awkwardness does not mean, however, that this approach was not used. To wit, Rolls Royce built a very successful Dart engine based on this principle. Today, the radial flow compressor is used in a number of smaller turboshaft engines in combination with the more promising and now much more common *axial flow compressor*.

The history of Whittle's efforts to build his first radial flow compressor engines with limited private and military resources is an interesting story. It speaks volumes of his tireless effort and his belief in the engine. It would indeed ultimately be accepted as potentially important for the military. As that was realized by the British military, his firm was incorporated into Rolls Royce, presumably for better management and access to the necessary financial and technical resources.

Von Ohain and Heinkel also tried to progress in developing better jet engines. While they enjoyed some success in garnering the interest of military funding early on, they eventually were beaten by other industrial concerns (Junkers and BMW) and succumbed to the needs of a war that was not going well.

Looking at the history of Whittle and von Ohain, it is possible to conclude that great ideas may come from individuals, but they soon lose control over them as powerful industry interests are drawn to the challenges and opportunities.

10.1 Axial Flow Compressor: The Bedrock of Modern Engines

A new compressor design approach was developed in Germany as well as in Great Britain. In Germany, the approach to finding the best compressor configuration for a jet engine was more deliberate and entertained a variety of approaches. The military establishments on both sides of the war were reluctant to spend funding and manpower resources on a new engine development program whose benefit might not be realized in time to affect the outcome of the war, then raging. Nevertheless, in Germany, two individuals in the Luftwaffe establishment managed to foster a funded program for a jet engine. The largest question was: how does one best build a compressor with good performance and good growth potential—within the context of a practical engine?

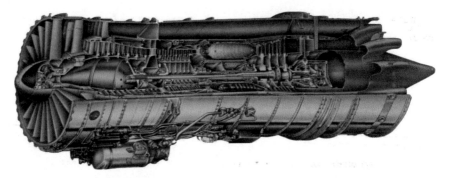

Fig. 10.2 Cross-section image of a P&W JT8D turbofan engine. The first set of blades at the flow entry at left is an inlet guide vane (Courtesy R. Lindlauf, New England Air Museum)

The best place to start an investigation about the inner workings of a jet engine is to examine one. Figure 10.2 is an image of a 1960s jet engine that is technically a fanjet engine (or turbofan) and it illustrates most of the attributes of both types of engines quite well. A fanjet contrasts with a *turbojet* in that the latter is not equipped with a fan. The first generation of jet engines were the simpler turbojets.[2] In this later turbofan, the fan is the first rotating set of blades that supplies air to a separate exit nozzle as well as to the engine's compressor (nearer the central rotor axis). The purpose of the fan is to increase the propulsive efficiency by increasing the air flow rate processed and reducing the average jet velocity. There were other advantages to the fanjet over the turbojet that led to their rapid adoption by the airline industry as soon as they became available: reduced noise and higher thrust at takeoff conditions when the necessary runway length was an issue.

We should note that the compressor and turbine in this particular engine are sectioned into two segments each, running on separate shafts rotating (one inside the other) at differing speeds. The components are referred to as high- and low-pressure compressor and with similar appellations for the turbine. The low-pressure shaft rotates inside the hollow shaft connecting the high-pressure components. The innovation of using two spools by Pratt & Whitney allowed the pressure ratio to be elevated significantly over a single shaft engine and it gave better off-design performance.

The speed(s) of the engine shaft(s) is one of the elements of engine performance displayed to the pilot of a jet airplane, usually expressed, not in terms of rotations per minute (RPM), but as a percentage of a nominal maximum value.

When this configuration was perfected,[3] it allowed for the building of two airplanes that had a profound impact on the jet engine industry, namely the US Air Force Boeing B-52 and the Boeing 707. The pressure ratio for this J57 engine compressor was around 12 (see Fig. 12.3) and for the later turbofan engine that

---

[2] A good image of a turbojet is shown in Fig. 12.3 where the illustration serves to illuminate aspects of a combustor design.

[3] In the P&W J57 or JT-3C, see Fig. 12.3.

employed the dual spool compressor concept illustrated in Fig. 10.2, it was almost 20. This turbofan powered several airliners built in the 1960s and 70s. A more modern GE90-115B turbofan engine (see Fig. 11.2) powering the Boeing 777 bettered this compression ratio with a value of over 40. The latest engine offerings by manufacturers in 2020 are 50:1 at Rolls Royce (Trent XWB) and 60:1 in (GE9X) in very efficient, very high bypass ratio engines.

To get a sense of the power modern engines process, at takeoff conditions, the turbine of the GE90-115B turbine supplies the fan with an estimated 120,000 hp and the compressor with more than 70,000 to produce 52 tons of thrust at takeoff processing almost one and a half tons of air each second! The power involved in the jet itself as the airplane sits at full power on the runway is difficult to estimate without access to proprietary data, but it can be safely said that it easily exceeds 100,000 hp. It is hardly surprising that when the gas turbine age dawned, builders of large airplanes never considered using the ICE again. The gas turbine engines made large, long range airplane possible.

10.2 Compressor Pressure Ratio

To this point, we have been rather vague about the importance of the compressor pressure ratio as a parameter in the design of a gas turbine engine. It must obviously be greater than one (no compressor), but how large? This is where the distinction between the piston (ICE or Otto[4] cycle) and the gas turbine (or Brayton cycle[5]) engines is starkly apparent. Fuel burning for the ICE is designed to burn out all the oxygen in the air to get the most heat and pressure from the process. In that engine, the compression pressure ratio (outlet to inlet) that the air is subjected to is strictly limited to what can be compressed without pre-ignition. In the compressor of the gas turbine air is not compressed with fuel as a fuel air-air mixture and there is no parallel limitation on compression. The compressor pressure ratio can be chosen to meet other criteria.

To be specific about this, we deal with two important temperatures that arise in connection with discussion of the gas turbine. These are the *turbine inlet temperature* and the *compressor exit temperature*. For brevity, these total temperatures will be called TIT and CET in this paragraph only. Heaven forbid we start speaking in engineering jargon! The CET is determined by the compression ratio as the compressor is close to isentropic. For example, the compressor outlet temperature from a compressor with pressure ratio of 40, operating in the stratosphere (- 60 °F, - 51 °C) is more than 800 °F (425 °C). This temperature will be even hotter if the

---

[4] The names Otto and Brayton refer to the individuals whose pioneering work in the development of these engines is rewarded with the name connections: the German Nikolaus Otto (1832–1891) and American George Brayton (1830–1892).

[5] The word "cycle" refers to a sequence of states that the air experiences in going through the engine and returning to the environment.

compressor is designed to have a higher value of the pressure ratio. The ultimate limit is reached when the CET reaches the maximum allowable TIT. At this design point little or no fuel can be injected and the oxygen available is plentiful. In such a design, the stoichiometry (the relative amounts of chemical reactants involved: fuel and oxygen) is fuel-lean, quite lean, in fact. However: no heat, no work. Evidently, there must be a middle ground between compression ratio one and the value that leads to the CET raised to equal the TIT. That is indeed the case. The virtues of an engine are two thermodynamic ones: work delivered per unit mass flow air processed (specific work) and work delivered per unit of heat provided (thermal efficiency). An engine with a high specific work will be compact and light in weight because the machinery involved will be appropriately modest while an engine design with high efficiency will be heavier but sparing of fuel use.

These two design conditions do not occur at the same compressor pressure ratio. In fact, as the pressure ratio is increased (in a design sense) the maximum specific work configuration is reached first, followed at a higher pressure by the maximum efficiency point. Thus, a choice for a preference must be made. The engineer would illustrate this with a cool graph, but we are not doing this here! (He would also have jargonized compressor pressure ratio as CPR!). The optimum pressure ratio has to be examined in the context of the question: how is the engine going to be used? Long distance flights require high efficiency for low fuel consumption while short flights benefit from not having to drag a heavy engine up to flight altitude and then back down, relatively often.

10.3 Compressor Aerodynamics

So how does the blading in these compressors work? The compressor is a machine that exploits the aerodynamics of wings. After all, the propeller is a compressor of sorts, why not build on that? Actually, the propeller is not a compressor, but it does accelerate a flow and the flow so produced could be slowed to raise pressure. A great advantage of flows in a compressor over similar flows in a freestream is that it has to be an *internal* flow with boundaries. That means that bound vorticity on the rotating and any stationary blading will largely avoid the creation of the trailing vortices that propeller blades must live with. A second advantage is that the flow behind such blading can be made relatively uniform, a virtue the propeller blade does not enjoy to the same degree. Finally, the turning that such blading imparts to the flow can be undone by stationary blades behind rotating ones. In order to do that, the compressor would have to be built with two kinds of blades, some that rotate and others that do not. Such a blading combination is called a *stage* and is illustrated in Fig. 10.2, when one looks closely. This arrangement is not unlike the two elements of a radial compressor with its a rotating impeller and a stationary diffuser. The same can be said about the blading combination of a counterrotating propeller.

Unfortunately, in order to fully understand how an axial flow compressor works, we have to draw diagrams that show the changes in speed and direction of the flow.

Fig. 10.3 Velocities and geometry of an inlet guide vane. The velocity in the direction of the engine axis is roughly conserved. In this figure and the several to follow, the leading edges are drawn as sharp. In practice they are rounded to allow for a variation in the in-flow angle of attack

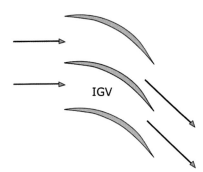

According to Bernoulli, speed changes that slow the flow result in the pressure increase we seek. A matter that is slightly complicating is that we will have to look at the situation in the reference frame of the blades themselves. In other words, looking at the rotor blades (the rotating ones) we have to look from a rotating viewpoint whereas the goings-on around the stator blades (the stationary ones) can be safely observed in the laboratory.

The flow through the compressor annulus (the space between the inner and outer cases) is helical in nature for reasons that will become clear further on. Commonly made representations of flow through such a compressor as straight-through parallel to the rotational axis involve an excessive use of artistic license and are inaccurate. Air that enters the compressor at 12 o'clock on the face may exit the compression process at 8 o'clock or thereabouts, depending on the design, specifically the compressor pressure ratio. For that reason, the descriptive word "axial" flow compressor is somewhat misleading, but we will stick to it because it is the accepted convention. To set up a helical flow pattern, the incoming axial flow will have to be turned by what is called an *inlet guide vane* (IGV). The first set of (stationary) blades on the engine in Fig. 10.2 are IGVs. This blade set turns the flow and necessarily accelerates it. The pressure will fall through this set of blades. That drop will be made up at the compressor exit where the last stator redirects the flow in the axial direction (in most design cases). Figure 10.3 shows the inlet guide vane flow velocities[6] where the axial direction speed is more or less conserved because mass flow rate has to be conserved. The "more or less" words have to do with flow area and density changes that may also be involved in such designs. In some compressor designs, the inlet guide vane is omitted. This omission will be clearer when we look at the fan in that family of engines.

The fan flow is the outer annulus part of the flow that is directed around most of the engine. The fan as a topic will be addressed again after we explore the basic elements of the engine. In the example engine shown in Fig. 10.2, the fan has a single rotor with an *exit guide vane* (visible at the entrance to the bypass duct) to align the flow direction to the rear without any rotating flow component.

The flow situation for the compressor stages that follow the inlet guide vane is illustrated in Fig. 10.4. The rotor turning as indicated accelerates the flow *away* from

---

[6] Velocity is *speed* with a specified *direction*.

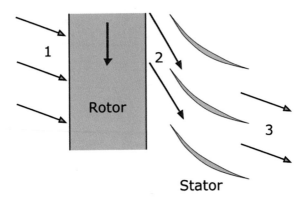

Fig. 10.4 Flow velocity vectors in the laboratory reference frame through a compressor stage. The axial components of all three vectors (not shown) are the mass flow determining flow velocities which are necessarily conserved. The rotor increases the flow velocity (from 1 to 2). The stator restores the flow direction (3) to approximately that of (1). The rotor details are illustrated in Fig. 10.5

the axial direction as shown. The rotor exit velocity in the laboratory reference frame at (2) is greater than at (1). The stator's function is to restore the angle of the flow (3) to roughly the orientation it had before entering the rotor (1). In its absence, the flow would be turned even further away from the axial by a following rotor, an impossible situation if more than a few rotors were employed. Without stators the result would be unacceptably high flow velocities and a limitation of the air throughput. As with a stator, the lengths of axial velocity vectors would also be conserved as with the inlet guide vane. The desired rise in pressure in the stator is realized because the incoming velocity (2) is larger than the outgoing (3).

How does the rotor increase the velocity as indicated? To visualize the velocities in the rotating reference frame, we must add the rotational velocity (red vectors in Fig. 10.5) and examine them in the rotating frame of reference. The green vectors (L1 and L2) in the laboratory reference frame are the same as those shown in Fig. 10.4 (1 and 2). The solid black arrows (R1 and R2) are the velocities in the rotor's frame. They must align with the geometry of the rotor blade at the leading and trailing edges. Proceeding through the rotor blades reduces the relative velocities and the pressure rises. This statement requires proof and fortunately one is available, but not

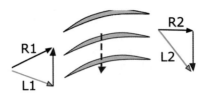

Fig. 10.5 Velocity vectors in the blade row of the rotor in its rotating reference frame. R1 is the relative velocity into the blade row and R2 is relative out. Green vectors (L1 and L2) are the laboratory reference frame velocities shown in Fig. 10.4 and the red lines are the rotational speeds for the change of reference

developed here to keep the narrative simple. In sum, pressure rises in both the rotor and the stator.

10.4 How Well?—Efficiency

The calculation of the pressure rise through a compressor blade row or stage is a simple application of the same energy equation used for determining the pressure on an airfoil, namely Bernoulli or its compressible equivalent. The velocities in play are those relative to the blades. That statement applies to both rotor and stator blading. The air state variables (temperature and density) will also rise as one proceeds from one stage to the next. The degree to which they change is easily related to the changing pressure through what is called a *stage adiabatic efficiency*. Such an efficiency relates the actual property changes to changes associated with an isentropic process, the accomplishment of which is always a design goal. While no one has ever been successful in devising a flow that perfect, the stage efficiency percentage is in practice quite high, reaching and exceeding 90%. It quantifies the modest frictional losses that occur. Its exact value for the compressor as a whole (the so-called *adiabatic compressor efficiency*) is of great interest to the engine performance evaluator. Its value can be gleaned from experimental testing or from detailed examination of the friction losses associated with the flow through the blading.

There is not much to talk about vortices in the flow through an axial flow compressor except that the boundary layers do contribute to rotational flow that ends up as turbulence. Rest assured, however, that the lifting airfoil behaves very much like the wing with a distributed vorticity distribution along the chords of the blades. The Kutta-Joukowski condition applies to fix the flow orientation at the trailing edge and the leading edges of the blades are rounded[7] to accommodate a variation in the in-flow conditions.

As with the wing, one can ask only so much of a blade row before the turning or angles of attack are too large for air to follow your geometric imposition on flow direction. The blading of a compressor can stall. Naturally, the compressor would want to be designed to be operated as close to stall as practical to realize as much compression per stage and to keep the number of stages to a minimum. As a matter of record, we note that early on in the history of axial compressor development, the pressure ratio per stage has grown from 1.15 (J47, 1947) to an estimated average of 1.28 (GE 90, 1995). This is no small feat because it dramatically reduced the size and weight of the compressor.

In practice, the number of stages in any compressor is significantly larger than the number of turbine stages because the compressor air moves into a region of

---

[7] In Germany, the need for rounding the leading edges of blades was recognized (in connection with development of the Junkers Jumo 004) with a patent application made in March 1943 by S. H. Decher. The patent titled "Geschwächte Schaufelprofilspitze," A.1808/E57/43 was granted in May 1944.

higher pressure. This adverse pressure gradient leads to the same flow separation issues associated with the wing aerodynamics. The turning that can be asked of each compressor blade is rather small and contrasts dramatically with the large turning angles in a turbine where the pressure falls. Considering that the power produced by the turbine is almost identical to that absorbed by the compressor, blading aerodynamics obviously plays a large role in determining the configuration of a gas turbine engine.

10.5 The Control Problem

The axial flow compressor operates with blades at an angle of attack to the air streaming through the compressor. The airstream speed and direction relative to the blades are controlled by the speed of rotation of the rotor blades and the engine's air throughput, i.e., the air axial velocity. If that combination of speeds leads to blade stall, the compressor becomes partially or totally nonfunctional. This possibility can occur when the pilot demands rapid changes in thrust. The pilot might therefore be confronted with the delicate operation of the compressor as well as that of the airplane. That is an untenable situation, especially in the military setting of the mid 1940s. An automatic control mechanism was and is necessary to shield the pilot from concern about the compressor operation.

Axial flow compressor development in Great Britain (by Metrovick) initially failed to manage such control and, in flight testing, the axial flow compressor was judged, at the time, to be unworthy of further development because of the excessive workload on the pilot. On the other side, a control system was devised by the German companies Junkers and BMW.[8] It was crude but effective. It consisted of an airflow rate control mechanism in the exhaust nozzle that, by measurements linked to it, always held the flow rate through the engine so that angles of attack on the compressor blades were always in a functional position and the compressor avoided stall.

It may be interesting to note that the German development effort initiated to find and produce a good jet engine involved a number of firms that include the two mentioned above as well as Heinkel (where Hans von Ohain was active) and Daimler Benz. The compressors considered included both radial and axial elements as well as hybrid radial/axial devices that spun the outflow, not in the radial direction, but rearward at roughly 45°. Most interesting is that Daimler Benz proposed a configuration that was to be a fanjet. This part of the history is described in greater detail in the bibliography reference entitled "Turbulent Journey" by the present author. When the war tide shifted against Germany, the scope of the jet engine development program was reduced to the projects by Junkers (Jumo 004) and BMW (BMW 003) with only the Junkers effort leading to a production program and the engine's use in primarily

---

[8] A patent (applied for in March 1942) was issued to S. H. Decher for this device in Germany. It was subsequently also granted in the US (after a postwar application) in September 1954 as no. 2688841 A.

two airplanes, the Me-262 fighter and the Ar-234 bomber. The Daimler recognition of the advantages of the turbofan engine was not singular. Back in the mid 1930s, Frank Whittle in Great Britain had recognized the advantages and patented the idea. The difficulties of developing the simpler turbojet were sufficiently challenging that he let the patent lapse.

That German control technology was shared with Allied victors who quickly recognized the value of the axial flow compressor and its advantages. Rather quickly, engineers learned to adapt more sophisticated control systems on the compressor. These successful steps ultimately led to the dominant use of the axial compressor in a large majority of engines. There were exceptions to be sure, but they were few and for special reasons.

Modern turbojet and fan engines employ electronic engine control systems to avoid stall. These systems manage the rate of power lever changes demanded by the pilot so that they are responsive and safe for compressor operation. A mechanical part of such systems may include variable stator angle control on the compressor so that flow angles are always close to optimal. In Fig. 10.4, the flow angle 2–3 is variable to suit such needs. Additional controls are necessary for starting gas turbine engines and these include bleed valves midway in the compressor.

The most common summary of performance of a compressor is on a so-called *compressor map*. It is somewhat akin to a drag polar of a wing and presents most of the important information regarding the characteristics of a compressor. It consists of a plot of pressure ratio versus air weight flow rate. Lines of constant rotational speed and constant efficiency are noted. The efficiency will have a maximum value at some point. Also shown is the line along which the compressor stalls. Stall is encountered at low air flow rates. The control system's function is to keep the point of operation in the region of highest efficiency and away from stall. Air weight flow rate and rotational speed (usually in RPM) are corrected to reflect inlet conditions to make the map as compact as possible. The corrections reflect inlet conditions that differ from those of a standard day, typically defined for aeronautical purposes as 29.92 in/Hg (14.7 psia or 101 kPa) and 15 °C (59 °F).

Chapter 11
Bypass and Other Engines

11.1 The Fan on a Turbofan Engine Has No IGV

Before leaving the subject of velocity diagrams in the compression parts of the engine, the reader may be curious about the flow through a fan that, in practice with high bypass engines, has no inlet guide vanes. The flow geometry is quite similar to that through a propeller as described in Chap. 8. The caption of Fig. 11.1 describes the diagram with flow that exits the stage in the axial direction imposed by the *exit guide vane*. A sketch of the flow through a fan of a modern turbofan engine is shown the figure. The comments about the helical nature of the flow through an "axial" flow compressor also apply to the fan, even with no inlet guide vane. The sketch shows the angular component of the rotor outflow in the laboratory reference frame (displayed as L2). The angular displacement of the stream tubes is much smaller than in a compressor because there is only one stage.

A couple of things about the engine in Fig. 11.2 are worth observing. First, the engine has some resemblance to a propeller driven system with the engine behind the fan as the power source. This and all high bypass engines have done away with the inlet guide vane and located the rotor to face the stream as the first blading element. Strong differences that set this engine type apart from the propeller with its ICE, however, are apparent. Among these is that the power source is the gas turbine in contrast to the ICE of yesteryear. That allowed power levels more than fifty times larger than what the largest reciprocating ICEs ever could. The fan is shrouded so that it can operate in an airstream that has been slowed by an inlet (not shown here) and there is an exit guide vane. Efficiency levels of this engine type are such that non-stop airliners can operate over distances covering half the globe. The post war ICE powered airliner was challenged to cross the Atlantic!

© The Author(s) 2022
R. Decher, *The Vortex and The Jet*,
https://doi.org/10.1007/978-981-16-8028-1_11

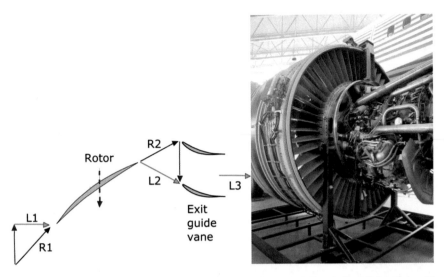

Fig. 11.1 Sketch: Flow velocity vectors through a fan without IGV. Flow is from left to right with vectors R1 and R2 relative to the moving blade. The laboratory frame rotor exit velocity (L2) is redirected in the axial direction (L3) by the exit guide vane. Image: Rear view of an engine fan shows the exit guide vanes (GE90 engine display at the Museum of Flight, Seattle)

Fig. 11.2 A modern turbofan engine: GE90-115 (mid 1990s) contrasted to a radial Wright 9-cylinder Whirlwind on a Ford Trimotor (1929); not to the same scale (Courtesy General Electric and agefotostock: Christopher L. Smith)

11.2 Bypass Ratio

The two fan jet engines pictured in Figs. 10.2 (1960) and 11.2 (1993) are embodiments of the goal to ever better propulsion efficiency by the reduction of the (average) jet velocity in favor of increased air flow rate. The amount of air bypassing the prime mover engine is expressed as a multiple of the air flow rate processed by the engine, i.e., the compressor, the combustor, and the turbine. Early in the age of turbofans the bypass ratio was around 1 (meaning fan and engine flow rates were about equal) and later 2. The widebody jets introduced bypass ratio of 5 and current airplanes may sport this ratio around 10. These increases in bypass ratio have very large impacts on fuel usage because of the improved propulsive efficiency.

The large fan in high bypass engines makes it imperative to consider its design for aerodynamic performance and, importantly, for weight. The early engines of this type, like the Pratt & Whitney JT9D powering the Boeing 747, had a large number (46) of high aspect ratio blades, meaning they had a relatively short chord. These blades had "clappers" between the blades to stiffen them and prevent harmonic vibrations from causing damage or failure. The blades were made of titanium alloys. Attempts to reduce the weight of the fan include the first design involving composite blades made with a metal leading edge to minimize foreign object damage by such events as bird strikes. These blades had a wider chord and were fewer in number. The General Electric GE90-115B has 22 such blades with composite materials (carbon fiber) at the rear of the blades (see Fig. 11.2). Another approach to weight reduction is the use of hollow all-metal, wide chord, blades. This was the approach taken by P&W in its upgrade of the JT9D (renumbered as PW 4000). The challenge of building a light-weight fan did not always go smoothly or without serious consequences, but it was met and serves well in modern fanjet engines. These blades are a tribute to the art and capabilities of modern aerodynamicists and engineers more generally.

11.3 Turboprops and Turboshaft Engines

The idea of the bypass turbofan is now fully implemented with fan engines used everywhere for commercial transport. An extension of the turbofan idea is the *turbo-prop* where nearly 'all' of the power is removed from the primary gas turbine engine flow to drive a gear box and with it a propeller. The effective bypass ratio is very large (larger than that of the ducted fan) but, as with an ICE powered airplane, the propeller sets an airplane speed limit.

Naturally, with an engine delivering its power by means of a rotating shaft (a *turboshaft* engine), all kinds of other applications can be imagined. Among them are helicopters, ships, etc. The success of the turboprop engines was very much due to their greater power output capability than were the piston ICEs. They also provided the user with engines that were much easier to maintain in service and could operate much longer between needs for service.

A few words about differences between gas turbine engines and the ICE will clarify the former's great advantage in aviation. The gas turbine is lighter as measured by power or thrust per pound of engine. Typically, at sea level conditions, 5–6 lbs of thrust or horsepower are produced by each pound of a gas turbine engine weight. This is quite an improvement over the ICE at 1 hp/lb where even that level is achieved only for fairly demanding circumstances such as those set for military aircraft.

The gas turbine uses a lower cost "jet fuel", similar to kerosene, rather than aviation gasoline required by the ICE. The lower volatility of the kerosene-like jet fuel also makes it safer.

Finally, we can say that the gas turbine with very few moving parts, and none of them reciprocating, does not require the mechanical maintenance of the ICE. To wit, the P&W R-4360 engine used in the Boeing Model 377 Stratocruiser airliner was fortunate if it lasted longer that 500 h between major overhauls. Allied bombers in WW II routinely replaced entire engines every 100 flight hours. These overhauls were costly in terms of labor and facilities and demanded an inventory of ready-to-go engines in case of an engine failure on an airplane that just came in. In commercial service, engine shutdowns were not uncommon, especially in the summer months. By contrast, the need for extensive overhauls of modern turbofan engines is in excess of 20 or 30 thousand (!) hours.

Chapter 12
Other Components of the Jet Engine

12.1 The Turbine

The turbine is the source of all the mechanical power in the engine. It can also be configured in radial or axial inflow configurations. An example of the former is similar to a child's pinwheel. Only Hans von Ohain used such a turbine in the first jet engine that flew in the Heinkel 178. Von Ohain and everybody else understood that the better way to go was to use an axial turbine, adapting the knowledge from steam turbines for power generation that had been in commercial service for over a decade. The radial turbine is now just a footnote, although one can find it used in automotive turbochargers where operating efficiency requirements are not as strict as they would be for an aircraft application (Fig. 12.1).

The description of the function of an axial turbine parallels that of the compressor. Rotors and stators are involved. The first element is stationary and is called a *nozzle*. This nozzle gives the flow kinetic energy in the azimuthal (rotation) direction which the rotor that follows it can harvest. The gas exiting the nozzle may be supersonic. The relative velocity incoming to the rotor will be lower than the exit velocity from the nozzle because the turbine rotor blades are receding away from it.

In contrast to the compressor where the flow in its stages is adequately describable as locally incompressible, the turbine flow is definitely in the compressible regime. The blading operates with a favorable pressure gradient so that large flow turning angles can be withstood without concern for stall. This aspect of the difference between compressors and turbine explains the physical difference between these components. Figures 10.2 (and 12.3) shows that the power transmitted from the turbine to the compressor is generated with relatively few turbine stages while the compressor absorbs the same power as input but must process the air gently (against an adverse pressure) to avoid the stall issue. To be specific, the P&W JT8D engine in Fig. 10.2 has a 1-stage fan and 13 (6 + 7) stages of compression, all driven by 4 (1 + 3) stages in the turbine. The contrast between compressor and turbine blading is starkly illustrated when we note that a single turbine blade row in the high-pressure turbine can power six stages of the compressor.

© The Author(s) 2022
R. Decher, *The Vortex and The Jet*,
https://doi.org/10.1007/978-981-16-8028-1_12

Fig. 12.1 A cutaway of an automotive turbocharger. Ambient air is drawn in from the left and exits in the spiral diffuser (blue). The red zone is the radial inflow (exhaust from the reciprocating engine) to the turbine wheel. It is relatively simple and performs adequately well (Photo: Quentin Schwinn (NASA))

The image of turbine nozzles and blades with their cooling air bleed holes (Fig. 12.2) suggests that the compressor feeds more than just the combustor. It also provides the cooling air for the turbine. The word "cooling" is a bit curious because the compressor air is, in fact, hot. It is, however, cooler than the combustor exit air and thus effective as a coolant. It must also be at the highest possible pressure to be able to enter the turbine flow through the many small holes. Identifying an optimum amount of cooling air from the compressor has to consider the positive aspect of operating at effectively higher turbine inlet temperature against the compressor power used to provide the cooling air. In modern engines, the cooling bleed is in the range of 5–15% of the air through the compressor.

Looking back at the ICE that also employs cooling air to maintain the temperatures of the cylinders, we note that the fuel energy transferred to the cooling air in that engine is lost for propulsion purposes. The gas turbine, by contrast, loses very little fuel energy heat as a pure loss. The mass of cooling air stays in the flow so that both heat and mass end up in the propulsion jet and contribute to the thrust produced.

12.2 The Combustor

We have largely left the topic of vortices to describe how a compressor and turbine work. To first order, vorticity is not a major issue in describing the workings of a jet

Fig. 12.2 Two blades of an aircraft gas turbine nozzle (one blade is covered by yellow tape) and rotor in the shop. This display is set up for training purposes. The rotor blades on the wheel to the left of the nozzle block are covered by orange plastic to protect the students. To the right of the nozzle element, the rotor blades are naked and turn the flow through a rather large angle. The single block element of the nozzle is held in place with a plastic tie. Note the heavy use of cooling air holes to bathe the nozzle material in cool air and minimize contact with the hot combustion gas. The second turbine stage is also shown as two stator sections (4 blades) (Courtesy General Electric, photo by author)

Fig. 12.3 A P&W J57 (or JT3C) in sectioned model form showing clearly the flow area expansion at the end of the compressor. The model illustrates the late 1950s technology of combustor design and the two compressor segments: low- and high-pressure elements powered by separate sections of the turbine with a connection by two shafts, inside one another. The artifact is located at the Wings Over the Rockies Air and Space Museum, Denver CO (cropped photo by Ryan Frost, https://com mons.wikimedia.org/wiki/File:Sectioned_Pratt_%26_Whitney_J57.jpg)

engine except to say that the bound vortices on the blades of both compressor and turbine must be in play. In the combustor, the vortex will be seen to play an important role. First, however, a look at the functionality of this component.

Combustion in a jet engine or better, a gas turbine, takes place at nearly constant pressure. The combustion process there takes place just as it does in a candle, the fireplace, or the burner of a hot air balloon. The consequence of burning fuel at constant pressure is that the air involved is heated and, being a gas, it expands proportionately in volume. The ideal gas law, $pv = RT$, is in play here. The resulting lower density ($= 1/v$) causes the heated air to rise in the fireplace chimney or, when contained by it, causes the hot air balloon to experience a net buoyancy and rise in the cooler atmosphere.

In reality, the (total and static) pressure in the combustion chamber of a jet engine falls a small amount for two reasons: pressure losses associated with the flow through the hardware and the process of adding heat to a flowing medium always leads to a loss in pressure. Any pressure drop is certainly to be avoided because the compressor had to provide it. From a technical viewpoint, one can show that this loss is minimized if the Mach number of the flow to which heat is added is kept small. This is done in a jet engine design by forcing the flow to slow with an area increase as it enters the chamber (see Figs. 12.3 and 12.4).

In the chamber itself, three major challenges are presented to the designer. These are, first, reducing the volume of the space necessary to carry out combustion to a minimum. Specifically, the length of the combustor must also allow the combustion

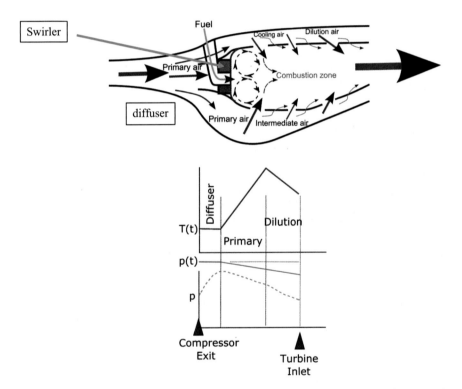

Fig. 12.4 Simplified sketch of a gas turbine combustor showing the swirler and other features and functions. The purpose of the swirler is to form a rotating mass of burning gas with helical vorticity on its outer edges. Simultaneously, the two illustrative rotational cells serve to continuously ignite the fresh air charge by bringing burning fuel forward (https://en.wikipedia.org: file: combustor diagram airflow.png by SidewinderX). The lower sketch is a simplified illustration of the changes in gas flow properties (total temperature in red and total pressure in solid green; static pressure in dashed green) that occur in a typical combustor that, in reality, is quite three-dimensional in nature

process to be complete. This is a problem because flame propagation speeds are very much lower than the flow speed through the chamber even when the Mach number there is held low. Here the vortex comes to the rescue. At the burner inlet one or more vortices are formed so that the swirling motion lengthens the residence time of the burning gas. This usually done by swirling the flow with an array that resembles an inlet guide vane (see Fig. 12.4a). The goal is to avoid forming soot, unburned hydrocarbons and simultaneously avoid forming pollutants such a carbon monoxide and nitrogen oxides. The second challenge is to produce a gas *uniformly* high enough in temperature acceptable to the turbine without itself being at risk of material failure due to overheating material surfaces or chemical oxidation. The uniform aspects have to do with the desire to avoid subjecting the blades to a variation in temperature that is hot and hotter at a high frequency. Lastly, the combustion must be stable in that it is, once lit, continuous and free of needing attention.

In practice, the burner of a gas turbine engine is operated as a combination of two processes: primary combustion where the chemistry is such that most oxygen is burned to achieve a stable and very hot flame and a secondary zone where additional air dilutes the combustion gas to a temperature that the turbine can tolerate. Figure 12.4 is a schematic representation of the process.

The lower part of the figure is a simplified sketch of the variation of the important flow parameters through the combustor. The red line shows a one-dimensional (averaged over the available flow area) variation of the total temperature, noting specifically the very hot primary combustion zone and the subsequent cooling by dilution with cooling air.

Also shown in the sketch is the variation of total and static pressures. The total pressure falls slightly during the process of heat addition to the air. The relative value of the static pressure is a reflection of the local Mach number. In diffusion, the total pressure stays constant while the static pressure rises because of the flow area increase. In the burner, the averaged flow behavior is for the Mach number to increase (toward thermal choking but not coming close to that) so that the net result is a decreasing (average) static pressure. The area decrease in the aft half of the combustor as shown in the upper sketch is an additional reason for the static pressure to decrease in the flow direction. While complicated, the design of the combustor mercifully involves subsonic, primarily low Mach number aerodynamics.

One cannot speak about the combustor without mentioning its pivotal role between the compressor and the turbine. The amount of fuel added to the flow must be just the right amount to raise the air temperature from that exiting the compressor to that acceptable by the turbine. The compressor air is hot from the compression work that went into it. Here is where the compression ratio of the compressor really matters.

In any functioning gas turbine engine, the amount of fuel added for combustion is limited and there is always more air processed than the amount necessary to burn the fuel. The excess oxygen flowing through the turbine has two consequences: the environment for the turbine is an oxidizing one that has the potential for setting the stage for material failure due to oxidation of the hardware in the burner and in the turbine. Good design learned from experience in the fields of metallurgy and cooling technique have largely put this concern to rest. The other consequence is that the

Fig. 12.5 Left: The voluminous combustion chamber on an early British engine, the Whittle W1 on display at the Smithsonian National Air and Space Museum. At right, the burner cans on a GE J47, an early (1947) single shaft, axial flow compressor engine (its turbine is at right under the plastic), on display at General Electric

left-over oxygen can be burned in an afterburner, i.e., in another combustor behind the turbine to obtain thrust for applications and situations where it is needed. An afterburner was used in the supersonic Concorde airliner to achieve cruising speed although not used during cruise. It is used primarily in military fighter jets.

Considering the progress made starting from the early engines of the 1940s to today's reliable machines, the goals in the combustor have largely been brought to a high degree of perfection. Included in the list of desired goals is the need to minimize undesirable combustion product emissions into the atmosphere. The burners have been so reduced in volume that they are hardly discernable as a component of the engine. Failures in operating engines are seldom attributable to the combustor, except perhaps when maintenance is given a short shrift. The combustion chambers and the region around the turbine entry are, therefore, the subject of scrutiny by the people who maintain jet engines.

Figure 12.5 also illustrates the need to keep the combustion chamber short. A long chamber will involve a long rotating shaft connecting the turbine to the compressor. If the shaft is insufficiently stiff in bending, it will be unstable. In the J47 of Fig. 12.5, engineers dealt with this problem by installing a bearing in the middle of the shaft.

In practice, the combustion chambers of a gas turbine are configured in one of two arrangements: straight-through flow or reverse flow. In most modern engines, the ability to achieve combustion in a small volume has led combustors to the use of the straight variety. The various engine cross-sections in this text illustrate this arrangement. Examples of the reverse flow configuration are shown in Figs. 10.2 (I-16) and 12.5 (Whittle W.1A). The reverse flow combustor allows for a short shaft between turbine and compressor.

12.3 Putting It Together into an Engine

How do the jet engine and its components work to produce thrust? We mentioned that the compressor and turbine function very efficiently. If the component efficiencies were an impossible 100%, then a compressor and turbine hooked together and could function without any consequence to the universe. They aren't, of course, so we have to pay for what they do. How could we do this? … and the answer is: by giving the turbine a higher *volume flow rate* to process. The only practical way to do that is to heat the gas prior to turbine entry so that it expands in volume—again, like the flue gas in the chimney.

We could start by adding just the right amount of heat to the air entering the turbine so that the losses in both components (compressor and turbine) are overcome and the engine operation is self-sustaining. That achievement probably brought the first reason for celebration to the pioneers Whittle and von Ohain! We could then add more heat. As the engine (of a fixed geometry) runs, the air mass flow rate through the engine must be conserved. It depends only on the product of flow velocity and air density. That means for the turbine that as the density is lowered by heating, the velocity must rise at every locale. In turn, the forces acting on the airfoils must increase because they depend on the *dynamic pressure* which depends on density and velocity squared. Heating the gas increases the forces, torque, and power from the turbine!

Whereas our ideal or self-sustaining operation left the outlet pressure behind the turbine at an atmospheric level, a greater heat supplied increases of the pressure at the turbine *outlet* because not all the pressure drop through it is needed to run the compressor. There is pressure left over. The engine becomes a pump: it processes air and delivers it at a higher pressure! With the addition of a nozzle, we can create a jet with (more) momentum to get thrust! Hot dog! More thrust is obtained by adding more heat up to the temperature limit that the turbine can tolerate.

The same physical argument about component entry temperature can be used (in reverse) to talk about the compressor where the work (i.e., the power) required for compression is also temperature dependent. Thus, if the compressor inlet air temperature is lowered, the power required is also lowered. One may conclude that the engine would best operated between temperature extremes that are as wide as possible. This is one reason that airplanes with jet engines operate so well in the cold stratosphere, where the temperature is around - 60 °F. Conversely there might be difficulties associated with operating an engine on very hot days when performance will be suboptimal.

The idea that wide temperature extremes are best for extracting mechanical work from a heat engine is a conclusion that the Frenchman Sadi Carnot (1796–1832) reached when he investigated the convertibility of heat to work generally. His examination of the performance of what we call heat engines laid the foundations for the Second Law of Thermodynamics that involved the identification of energy (through the First Law) and entropy as properties, the concept of reversibility, the notion of temperature, the role of the temperature extremes that apply to heat engines and limit their performance. His studies launched a great deal of understanding of the workings of nature and that knowledge fueled an important second period of the Industrial Revolution. The first having been centered on the use of the steam engine.

A truth that follows from the work of Carnot and others is that a perfect heat engine (a Carnot engine) can be imagined, although building it is not practical and may not be possible. The gas turbine engine, like all engines, is limited by the temperature extremes available. Specifically, the successful generation of mechanical power by means of any *heat engine* depends on heat from a high temperature source (or reservoir, as thermodynamicists like to say) intercepted and converted as it flows to a low temperature reservoir (a heat sink). Without *both* of these reservoirs, nothing useful can be made to happen. Consider, for example, that the heat engine that is planet earth and the life on it operates between input from the sun (~ 5800 K or some large number of degrees F) and radiates the waste energy to the near zero absolute temperature of space (~ 3 K). Consider also that a uniformly hot environment is not a place where much can happen. Is that where the notion of 'hell' being hot originated?

Enough of philosophical wanderings and let's get back to our gas turbine engine! The arguments made above concerning the working of a gas turbine engine can be made much more precise using an expanded version of the conservation of energy statement we have used so far to include mechanical work (compressor and turbine) and heat (combustor). The books referenced in the bibliography should be helpful in the quantification of the performance of the engine. The analysis is rather simple in that the relations are algebraic and don't involve much higher mathematics. The takeaway from such analysis is that the temperature extremes involving the maximum (turbine inlet) and minimum (compressor inlet) play a dominant role in the performance.

Finally, is this new engine any good? It is, apparently, a very successful engine to propel airplanes. As a postscript to our description of the gas turbine, it merits callout to some of the features that make it different from the ICE we know so well. The jet engine is a thrust, rather than a power, producing engine. Both engine types use liquid fuels, albeit slightly different kinds. For flight, using a *liquid* fuel is extremely important because of the ease of storing it onboard and handling it. Just contemplate for a moment the need for shoveling a solid fuel like coal on the ships of yesteryear or the steam locomotive. The difficulties involved in storing gaseous fuel under pressure in an airplane where payload volume is very dear have been considered and are not attractive. Aviation had to wait for the invention of the ICE, in part, for that reason.

There are important technical differences between the engines that say a lot about where and how they are used. The ICE works well and efficiently at power levels less than maximum. The gas turbine does not. The gas turbine works most efficiently at full thrust (or power) which is just what is needed for an airplane in cruise. The ICE works very well in automobiles where it hardly ever runs at full power. The demands of traffic insist on that.

In considering the turboshaft gas turbine for small applications, say under 500 hp, (about 400 kW) the boundary layers on the walls of the various flow components become a larger proportion of the total flow, meaning that friction is proportionally more important. The associated reduction in component efficiencies and the total pressure losses become important contributors to increased fuel consumption (per horsepower). The gas turbine is not easily well adapted in small sizes with high efficiency as a goal.

Finally, an important operational difference is that the turbomachinery, the compressor and turbine, has a large inertia invested in rotation. That will result in a slower response to power demand changes when compared to a similar requirement made of the ICE. The slower temporal response has consequences in the military setting or when emergencies are encountered.

12.4 How Do You Start This Thing?

When we examined the ideal engine consisting of perfect (thermodynamically reversible) compressor and turbine, we implied that all we had to do is spin it up to operating speed and then let it run on its own. That is indeed what has to be done to a real engine. Gas turbine engines typically have connections to a device that has enough power to raise the speed to the point where the engine operation is self-sustaining. The power for an electric motor may be provided by a battery, an auxiliary power unit (an APU), or from a power cart on the ground. Typically, commercial airliner engines are started by an air turbine supplied by an APU. Another source could be air from a storage source or, as it has been done, generated by burning a charge of solid rocket propellant-like material to generate the gas flow. There are interesting circumstances under which such means were or are used but they are quite uncommon. In flight, an engine may "flame out" or stop working and needs to be restarted. In some circumstances, the airplane's speed and altitude may be exploited to have the flight airstream turn the engine over. It is hard to imagine that the circumstances that call for such action are benign.

12.5 Bleed Valves and Variable Stator Geometry

There is another aspect of starting a jet engine that is interesting. Consider the operation of an axial compressor. For a running situation, the blade height and flow passage widths are very much smaller at the outlet end of the compressor than at its inlet. The reason is that the increasing density associated increasing pressure as we look further down the compressor path is such that a smaller flow area is required at the high-pressure end to pass the mass flow being processed. During starting, that high density is not yet achieved but the mass flow forced in by the inlet must exit at the outlet. That implies that, initially, the flow velocities near the rear are very high, potentially high enough to "choke" the flow by reaching sonic speeds. If and when that happens the increasing flow demanded by the front of the compressor cannot be accommodated and the process is stuck; unless…

Fig. 12.6 External view of a small gas turbine compressor with variable geometry stators. The five rings are attached to levers that change the orientation of the stators mounted on shafts through the case. The hydraulic lines and the associated cylinder are used to fix the stator orientation

There are two aspects of the design of a modern gas turbine engine that are used to prevent this situation from being problematic. The first is to allow a bleed from the midsection of a compressor to remove partially compressed air from that location so that flow to the rear blockage is reduced. These bleed valves shut when operation reaches a normal operating state.

A second aspect that deals with starting is that in some compressor designs, the stators can be made to change the angles they present to the flow. Thus, they can be oriented to minimize blockage during startup. These two solutions can be employed in concert when they are designed to do so. The pilot sitting in the cockpit's left seat is ignorant of all this because there is an engine control system that takes the necessary measurements and, with actuators of various kinds, does all that is necessary without burdening the pilot with its doings (Fig. 12.6).

Finally, the starting process is eased by the configuration of the engine when it has more than one shaft. Specifically, the starting process is initiated with the high-pressure components, the high-pressure compressor, burner and turbine, while the low-pressure components wait for things to happen on the inner portion of the engine. After all, the high-pressure section of a multi-spool gas turbine is an engine within an engine! When the high-pressure section is running in a sustained way, the low-pressure components follow suit and the entirety springs to life.

Chapter 13
More Components: Inlets, Mixers, and Nozzles

13.1 Inlets

We return to vortices to see where they play important roles in the operation of a jet engine. One aspect of an engine installation design is reflective of the comments made on the wing design. Recall that a nice rounded leading edge on the wing is desirable to allow for operation over a wide range of angles of attack. That same thinking also applies to the inlet that has to handle air nicely in flight and on the ground where the airplane is almost stationary. At low speed, the stagnation point (line) on the inlet air flow is located on the outside of the nacelle. Near a nacelle's maximum width, the curious person will find a red line and words to the effect: "don't stand forward of this line during engine runups on the ground, or else"! The point is to warn persons not to participate in the suction exercised by the engine when it is running on the ground. The roundedness on the inlet parallels the design of a wing at its leading edge. The inlet must manage inlet flow to be uniform in low-speed flight, specifically at high angle of attack or with a cross wind on the runway. The rounded lip allows the entering air flow entry to be relatively uniform. At cruise speed the stagnation line is almost coincident with the foremost *highlight* of the nacelle, its leading edge if you will. The two stagnation streamlines are shown in Fig. 13.1.

13.1.1 An Old Wives' Tale About Inlets

Descriptions of how a jet engine works are sometimes said to involve the notion that the engine "sucks" itself along to propel the airplane. The reality for an airplane in flight is that nothing could be further from what actually happens. In flight, the purpose of the inlet is to slow the freestream. In this way, the inlet performs part of the compression process. There is no suction involved in its function in a cruise flight condition. The question really has to be centered on the pressure at the engine

R. Decher, *The Vortex and The Jet*,
https://doi.org/10.1007/978-981-16-8028-1_13

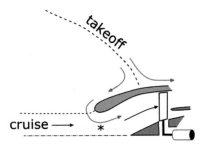

Fig. 13.1 Sketch of the flow pattern in and around a subsonic airplane inlet at low speed on takeoff (green streamlines) and at cruise (red). Dashed lines are the stagnation streamlines. The * denotes the inlet throat area. Under all conditions, the flow Mach number at the engine face is about 0.5

inlet face. That, in turn, has to do with the flow speed, i.e., the Mach number in that location.

For the sake of argument, one can say that the flow Mach number at the engine face, be that a fanjet or a turbojet is about 0.5. The airflow around the blades requires that under most conditions. Subsonic flight inlet performance is close to ideal, meaning the effect of total pressure losses are minimal. One can say that the flow from the outside world where the pressure is the *static pressure* to the engine face always preserves the total pressure associated with the flight speed.

At takeoff conditions, the static pressure at the engine face (local $M = 0.5$) is about 20% lower than the outside (where the local Mach number is near zero), hence air is indeed sucked in. However, during flight, say at $M = 0.8$, the conservation of total pressure will *increase* the static pressure at the engine face because the inlet *decreases* the flow speed. Under such conditions, the pressure at the engine face is about 30% *higher*—no suction here! At flight conditions like those of a Concorde ($M = 2$) or the SR-71 ($M = 3.2$), the pressures at the engine face are about 7 or 40 (respectively) times the static pressure outside, even though these inlets suffer some (minimized by design) loss in total pressure because of the shock waves in the inlet. Certainly, no suction there either! On a supersonic airplane, the inlet does a significant portion of the engine's necessary compression. On a ramjet engine that is necessarily flown at supersonic speeds, the inlet does *all* the air compression and the use of rotating machinery is dispensed with altogether in such an engine.

The proper way to attribute performance contribution by the inlet relative to the engine is to discuss the amount of compression done by the inlet and compare that to the amount done by the compressor. One cannot talk about the relative contributions to thrust because inlet, engine, and nozzle work together as a system. It would be a little like talking about the contribution to a runner's athletic performance by his liver!

In flight, the higher pressures on the inlet do raise hardware design issues. The associated forces have to be taken into account for the mounting hardware that is used to affix the inlet to the engine. Insufficiently strong bolts will result in a forward departure of the inlet and the likely end of proper engine operation. The same pressure

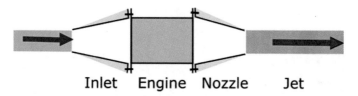

Inlet   Engine   Nozzle   Jet

Fig. 13.2   Forces on ancillary engine components. At the engine faces (front and rear) the pressures are higher than outside, hence the bolts holding the inlet and nozzle onto the engine are in tension

forces also play similar roles at the nozzle. High inside pressure and low outside pressure there also require a strong connection between nozzle and engine. The forces associated with that of the restraining bolts have nothing to say about engine thrust (Fig. 13.2).

The DC-10 inlet in Fig. 13.3 is of a typical design for commercial airliner. The internal surfaces of the inlet handily serve as a sound absorbing liner to reduce noise emissions. The many small holes and the cavities behind them are designed to absorb the noise frequencies associated with the blading behind the inlet. The figure also illustrates the use of a vortex generator on the shoulders of the nacelle to improve the flow over the wing that might have stall issues due to the nacelle 'shadowing' the wing at high angles of attack. This vortex generator (there may be two of them in some applications, humorously referred to as "rabbit ears" but technically called a 'strake') is very effective at low speed and aerodynamically close to absent in cruise when the external flow aligns itself closely with that of the 'ears'. In some instances,

Fig. 13.3   The inlet of a DC-10 airliner wing mounted engine nacelle. One of the two "rabbit ear" vortex generators is located at the 11 o'clock position of the cowling

Fig. 13.4 A LAN (Argentina) Boeing 767 taking off from Los Angeles. The white "cloud" just inside the inlet is water vapor condensation at the minimum flow area where static temperatures are lowest and, in this case, below the dew point (Photo courtesy Werner Horvath, Airliners.net)

the strake may also be made effective at improving the flow in the space on either side of the strut and under the wing.

On a subsonic inlet like that shown in Fig. 13.3, the highest local flow velocity is always close to the location where the flow area is a minimum. This area is called the *throat*. The flow velocity there is close to sonic but nominally subsonic. During operation at takeoff, the velocities there are high enough that the local static air temperature can drop below the dew point so that condensation takes place, just as it may on the upper surface of the wing. Figure 13.4 shows an airliner taking off under such conditions. There is little performance impact by this condensation!

13.2 Inlet Geometry and Diverters

On airplanes capable of supersonic flight, the inlet is more complicated. Total pressure recovery, i.e., minimal loss of the total pressure available, is an important design goal. This is a concern from two aspects: total pressure losses through shock waves and flow uniformity if the inlet is located along the fuselage of the airplane. The inlet to a $M = 2.2$ airplane (Fig. 13.5, a McDonnell F-4) embodies features that deal with these concerns.

Fig. 13.5 A McDonnell F-4 in flight showing the shock generation ramp on the inlet and the boundary layer diverter between the inlet and the fuselage (Picture credit: USAF—holloman.af.mil, curid = 50920146). At right is the Museum of Flight display showing the diverter head-on. Note the small diverter on the scoop at lower right and the inlet instrument probe

The first order is to minimize the losses associated with shock waves. Total pressure losses are minimized when the flow can be made to slow through a series of oblique shock waves rather than a single normal[1] shock. The flow entering the engines may or may not have gone through the shock wave generated by the nose of the airplane. Typically, a wave is created by a ramp that reduces the flow Mach number by turning the air flow slightly outward. On each engine, the F-4 has such a flow deflecting ramp adjacent to the fuselage. In general, the inlet compression process is improved when the flow is turned by more than one ramp. This is usually employed for airplanes where operating efficiency is at a premium. The Concorde is an example, Fig. 13.6. In that example, the shock generated on the underside of the wing is the first wave encountered by the airflow to the engines.

In the F-4 image, the ramp is not on the body but offset by a few inches. That offset is to address the second aspect of the design of a supersonic inlet: avoid ingestion of the air slowed by the boundary layer on the airplane's body.

The inlet of the Concorde supersonic airliner is illustrated in a takeoff configuration (photo) and in cruise flight condition (sketch) of Fig. 13.6. There are three wave generating ramps with two hinged elements for each inlet. In the photograph, these are fully retracted so that maximum airflow is admitted as, for example, during takeoff. The keen observer will note doors on the underside of the inlet to discharge excess air or admit additional air when necessary, notably during takeoff. The sketch shows schematically the array of oblique shocks generated by the three deflection surfaces in a cruise configuration. A normal shock will conclude the diffusion process of the supersonic flow. Further slowing of the air takes place in a diverging duct much like that in a subsonic transport engine.

---

[1] A normal shock reduces a supersonic flow to subsonic in one (irreversible) step, see the shadowgraph in Fig. 7.5.

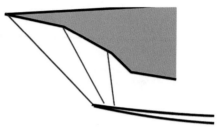

Fig. 13.6 The two-dimensional inlet of the Concorde supersonic airliner. The wave generating ramps (there are two hinged elements for each inlet) are fully retracted in this picture so that maximum airflow is admitted as, for example, during takeoff. The ramp has the word "DANGER" noted. Note also the diversion of the boundary layer air between the top of the inlet and the wing. The boundary layer is thicker on the inboard side of the inlet, hence a wider gap. The boundary layer air from the underside of the delta wing is dumped to the side of the inlets

The geometric variability together with the need to control the shock boundary layer interactions required for an inlet designed for supersonic flight is very complicated and involves a costly part of the design process of the propulsion system. This description of the essential features of such inlets is necessarily quite simplified. The reader may wish to consult more detailed descriptions of such inlets, specifically those for supersonic military bombers.

The inlet on a Lockheed F-104 ($M = 2$) employs (half) an axisymmetric (conical) shock wave to slow the flow (Fig. 13.7). The boundary layer air is also diverted on

Fig. 13.7 Axisymmetric inlets for supersonic aircraft, F-104 and SR-71. In the lower right background of the SR-71 is a display of a P&W R4360 with its 28 cylinders (Photos taken by the author at the Museum of Flight in Seattle)

that airplane. A good example of a whole axisymmetric inlet is that of the SR-71 ($M = 3.2$) where a conical shock wave is generated on the sharply pointed cone. From there the air enters the internal portion of the inlet and proceeds to encounter a number of relatively weak shock waves. This inlet is also designed to remove the boundary layer air on the internal inlet surfaces by suction on these surfaces to minimize the deleterious effects of shock/boundary layer interactions. These can cause flow separation from the wall much like those discussed in connection with transonic drag rise for a wing.

In all cases, inlets will complete the diffusion process with a divergent flow section where the near sonic flow is slowed to near $M = 0.5$ that the engine's compressor blades can handle.

13.3 Mixing

Internal flow mixing in an engine was carried out in many turbofan engine configurations of modest bypass. The Pratt & Whitney JT8D shown in Fig. 10.2 is a good example. The mixer can be seen at the rear of the engine where fan air and engine core exhaust are mixed. The idea is to create a single, uniform jet stream for expansion by a common nozzle. This design is a follow-on from the earlier P&W J57 engine-based JT3D low bypass engine where the fan flow was directed out of the engine via a separate exit nozzle. The performance of mixed and separate flow configurations is quite similar, except that the mixed flow engine has lower noise emissions. A view of a mixer for a high bypass engine from the rear is shown in Fig. 13.8. The idea is to create a large contact area between the two flows with a wide vortex sheet between them so that mixing takes place rapidly. The differential speeds of the mixing flows generate the vortex sheet that allows the mixing.

Modern high-bypass and very-high-bypass engines may or may not sport fan/primary flow mixers as the flow rates are quite disparate. The performance advantage gained has to be balanced against the weight penalty of the associated hardware. An example of a separate flow high-bypass nozzle configuration is shown in Fig. 13.10.

A substantial amount of mixing of very hot and cooler flows also takes place in the combustor to achieve acceptable turbine inlet temperatures as described in Chap. 12.

13.4 The Nozzle

The last element of a jet engine is the nozzle where the high-pressure gas produced by the pump that is the engine, expands to atmospheric pressure and in the process, creates the propulsive jet. In its simplest form, a nozzle is a duct of decreasing flow area (a convergent nozzle). This geometry is universally used for turbofan engines where the flow exits the engine with a pressure that is not sufficient to drive the

Fig. 13.8 An internal mixer for engine core flow emanating from "behind" the round central cone with the fan flow originating in the surrounding darker space (Photo by author at the service education facility of General Electric in Evendale, Ohio)

jet to sonic speed. A ratio of total to static pressure (outside) on the order of 1.9 is required to do that. In engines for commercial airliners, the *nozzle pressure ratio*, as this number is called, is kept to lower values because of the need to avoid the wave generation by supersonic turbulent parts of the flow. This aspect of the turbulent mixing with the freestream is the source of much jet noise in high-speed jets. In fact, in the early *turbojet* transports, the supersonic flow noise was so severe that extraordinary means had to the employed to minimize it. The approach used then was to force rapid mixing with the freestream air by external *daisy petal mixers* (Fig. 13.9) and similar devices. Their use imposed a thrust and fuel consumption penalty that had to be borne. The advent of the turbofan obviated the use of such mixers because the jet total pressure was significantly lower.

The older reader may remember listening to the roar of a 1970s commercial jet taking off. The crackle in the tonal mix was from the supersonic cells interacting with the environment. Some business jets may still be equipped with engines with the higher nozzle pressure ratio and their noise may be enjoyed by those who wish to do so.

External mixing is also done on modern jet engines as, for example, on the Boeing 737 shown in Fig. 13.10. The goal is also to limit the volume of the region where the mixing takes place that is the source of jet noise. This is done by getting the mixing done quickly. The scallops in the trailing edge of the nozzle could also be properly described as vortex generators.

The core flow of the engine shown in Fig. 13.10 expands the jet by means of an *external expansion nozzle* that is visible as the central cone of that engine (no

Fig. 13.9 An early turbofan engine on a Boeing 707 (a Rolls-Royce Conway) with a daisy petal mixer that joins the external and the primary core flow of the engine. In this engine, the fan flow exits through two nozzles on the sides of the nacelle. These nozzles are partially obscured in this view but the visible one exits the nacelle between 2 and 4 o'clock positions (The Boeing Company)

Fig. 13.10 The (convergent and scalloped) fan flow nozzle exit lip of a commercial airliner, a Boeing 737 MAX. Both core flow and fan flows are partially expanded against the conical, external nozzles (Photo by author at the Boeing assembly plant in Renton, Washington)

scallops on the exit lip). An external expansion nozzle operates pretty much as does the axisymmetric inlet, except in reverse: The nozzle accelerates the flow while the diffuser (or inlet) decelerates it. Both devices handle flows that are fairly close to reversible (free of losses). The fan flow in Fig. 13.10 also expands against an external nozzle wall, but to a lesser degree.

This discussion of the nozzle and that of the inlet allows a revisit to the subject of property variations through the entire engine. Specifically, the sketch of Fig. 9.2 can be revisited to show the variation of temperatures and pressures through the jet engine as a system. Figure 13.11 is the earlier figure augmented with the variation of the gas properties through the inlet and the nozzle. The solid lines are of the total values (in the engine reference frame) with the dashed (red and green) lines are an indication of the static values. The static pressure varies from atmospheric value back to atmospheric value, while the higher static temperature in the jet reflects the need to comply with the Second Law of Thermodynamics that requires a certain amount of heat to be wasted. For purposes of this sketch, we consider a simple turbojet rather than a turbofan where two flows would be involved, making for an overly complicated sketch. The variation of the fan flow can be inferred from that through the core engine, without heat addition and without a turbine. Further, to simplify matters in the sketch, the inlet and nozzle flows are taken to be reversible so that the total pressure losses associated in these components are overlooked. In reality, especially in the inlets of supersonic flight airplanes, some total pressure losses would be experienced.

A textbook on propulsion nozzles will state that thrust from a nozzle consists of two parts: the momentum of the moving gas and a force resulting from pressure mismatch between atmospheric pressure and the pressure at the nozzle exit plane. The pressure mismatch contribution to thrust is small in jet engines and absent when the jet flow is subsonic because the pressures in question necessarily equilibrate.

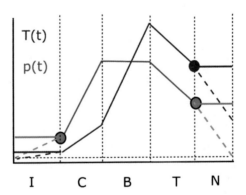

Fig. 13.11 Schematic variation of properties through a turbojet with reversible inlet and nozzle flows. The black dotted line establishes the level of static quantities in the air processed by the engine. The solid lines are total quantities in the engine reference frame. The dashed lines describe the variation of static quantities. The letters I … N refer to inlet, compressor, burner, turbine, and nozzle, respectively

Hence, we can speak of thrust being largely the momentum term for such nozzles. For propulsion systems in high performance aircraft, the pressure mismatch term does contribute a correction that can be minimized by proper design, but it is safe to say that the momentum term in thrust dominates the performance so that this aspect of thrust can be relegated to being a detail addressed when performance accuracy is desired. For rocket engines designed to fly into space the matter of pressure mismatch is significant as we shall see. For the present, let us stay with jet engines.

13.5 Choking

For military jets that need large amounts of thrust and fly much faster than transonic commercial transports, the nozzles are often designed and operated with a sonic flow in the throat of the nozzle and further expansion in a divergent cone. Such nozzles are called *convergent-divergent*. The jet from such nozzles is supersonic with a high speed of sound because the gas is hot and it is consequently noisy. Military operations are not as concerned with noise as flight operations at civilian airports are. Often, the engines of military aircraft are equipped with afterburners. When such an afterburner is in use, the exhaust gas is even hotter and lower in density so that the flow limiting throat has to be enlarged to pass the air volume that the engine provides. For this reason, afterburner engines usually have variable geometry nozzles, specifically the throat opening size than can be altered as conditions require.

When the afterburner is in operation, the Mach number of the hot flow in the duct where combustion takes place tends toward $M = 1$ as heat is released by the burning fuel. Under such conditions a limitation on flow rate may be imposed by *thermal choking*[2] rather than *geometric choking* set by the minimum flow area or throat. A geometric throat minimum is hardly discernable in Fig. 13.12 on the B-58 engine nozzle displayed in afterburn mode which means that the process of afterburning drives the flow Mach number quite close to unity.

Flow choking is a very important aspect of internal flows, i.e., flows through engines. When a flow is choked, the mass flow rate through the area where $M = 1.0$ is a maximum that can only be influenced by upstream total pressure and total temperature because these two quantities determine the mass density of the gas and the local speed of sound. Specifically, the mass flow rate cannot be altered by lowering downstream pressure because information about the pressure there cannot be transmitted upstream. In practice, choking is a controlling factor in the ability of internal combustion engine's ability to process air past the valves. It is also a factor in controlling the flow through the gas turbine engine because the turbine nozzle is

---

[2] Choking is the phenomenon where flow rate in a stream tube is governed only by conditions upstream of the location with choked flow, i.e., the total pressure and temperature, and not by the downstream static pressure.

Fig. 13.12 Left: The propulsion nozzle of a supersonic B-58 bomber. Note the throat and the divergent cone that is variable in geometry (in both throat and exit areas) to accommodate afterburning. The flame-holders for this device are visible at the far end of the afterburner cavity with enough flow length to the nozzle throat to allow for more-or-less complete combustion of the fuel added (Photo by author at the National Museum of the USAF). Right: On the Space Shuttle 'Discovery' at the Udvar-Hazy Center of the Smithsonian National Air and Space Museum, the nozzle throats are near the bulkhead and are roughly one eighth the nozzle exit skirt diameter

always choked whereas the primary propulsion nozzle of a jet engine may or may not be. Finally, choking is a central issue in flow through the rocket nozzle.

13.6   More Extreme Nozzles: The Rocket Engine

The requirement for a nozzle area minimum in an air-breathing engine nozzle parallels that for a convergent-divergent rocket engine where nozzle pressure ratios (chamber to ambient) are much greater and the exit to throat area relation is much more pronounced. The three hydrogen–oxygen rockets engines ('acronymed' as the SSME, the Space Shuttle Main Engines) on the Space Shuttle illustrate this (Fig. 13.12). The steady pressure in the combustion chamber of these engines is more than two hundred times atmospheric pressure (about 3000 psi). During its ascent flight, the nozzle pressure ratio varies from about 200 at sea level and rises dramatically during ascent. Such high nozzle pressure ratios require a significant divergent nozzle skirt so that as much of the available thrust can be realized by matching the flow pressure at the nozzle exit to the atmospheric pressure. The actual design increased the throat area by a factor of 69 and forced the flow to a jet Mach number of about 5 at the nozzle exit.

Fig. 13.13 The over-expanded SSME on a test stand running with oxygen and hydrogen. The absence of hydrocarbons makes for a beautiful and clear image of the flow (NASA)

13.6.1   Overexpansion

At launch or as illustrated on a test stand (Fig. 13.13), the nozzle is *overexpanded*. The nozzle skirt expands the flow too much, i.e., it is too long and should have been taken to only about 15 times the throat area. At this condition, the flow must recompress after leaving the exit to match the high atmospheric pressure. This can be seen in the converging boundary of the jet and in the observation of a normal shock wave to bring the flow closer to the relatively high pressure of the environment. Not visible in the picture (because the jet gas is nearly transparent) is that the first (weak) compression wave is a conical oblique shock wave that connects the exit lip of the nozzle and the perimeter of the central normal shock in the image. The normal shock doing the recompression in the middle of the stream is evidently strong because of the greater luminosity behind the shock resulting from the higher static pressure and static temperature (reconversion of kinetic energy to heat!). These static conditions would be those measured by an unfortunate (theoretical) bug were it to make this trip through the nozzle!!

13.6.2   Under-Expansion

As the vehicle ascends in the atmosphere, the atmospheric pressure around the vehicle and engine falls, and the need for recompression is diminished. The normal shock seen in Fig. 13.13 will move to the rear and weaken. There will be an altitude where the expansion is just right and matches the flow pressure at the nozzle exit. The nozzle will be perfectly expanded at this point. For the SSME engines, this is estimated to occur at an altitude of about 15,000 feet. As the vehicle ascends further, the (static) pressure at the exit will stay as it was and the flow will have to expand further outside to match the ever lower atmospheric pressure. The nozzle is now *under-expanded*. A rocket nozzle design for ascent to space must be optimized for this aspect of performance and the important consideration is the engine weight, specifically the weight of the nozzle skirt.

The nozzle pressure ratios when the engines approach shutoff during flight are very large. Near the end of burn, the expansion after the flow leaves the nozzles is dramatic as shown in Fig. 13.14. The picture is of the Saturn V launch of Apollo 11. Such a nozzle is under-expanded in this picture. Ideally one would have wanted the nozzle skirt to be as large as the expanded jet in the image, but that is evidently not realistic. Here, the jet is made visible because the combustion is of a hydrocarbon fuel rather than hydrogen. In a similar view of the Space Shuttle at high altitude and after solid rocket booster burnout, the large external jet would be hard to see because the water vapor (and a substantial amount of hydrogen) jet produced is almost transparent to visible light.

Fig. 13.14 The Saturn V rocket in flight carrying Apollo 11. Note the very large width of the jet from an under-expanded nozzle (NASA)

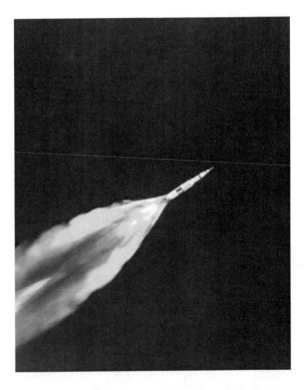

Fig. 13.15 Over- and under-expansion waves in an afterburning J58 propelling an SR-71 (NASA)

The conical waves that start at the exit lip (Fig. 13.13), both compression shock or expansion waves, are the beginnings of a train of waves that interact with the free boundary and reflect from it in the opposite sense that they impact the boundary. That is, a compression wave will be reflected as an expansion wave and vice versa ad infinitum. Thus, one can imagine the succession of waves in the jet getting weaker as one proceeds further from the end of the nozzle. While it is hard to see this in the image of the Saturn V flight, it is easily observed in an afterburning turbojet (see Fig. 13.15). The picture is of an afterburning Pratt & Whitney J58 on an SR-71 in flight.

The alternating nature of reflected waves by the free boundary of a jet contrasts to the reflection of waves in a flow with rigid boundaries. There, waves are reflected in kind, meaning that a compression wave is reflected as another compression wave. To some extent this is exploited in the design of high Mach number supersonic inlets where supersonic flow enters the inlet and the internal shape of the inlet causes a series of reflected compression waves to slow the flow as isentropically as practical. Often, although not necessarily, such inlets use an external cone to reduce the flow Mach number from its flight value to something lower where the internal inlet flow completes the Mach number transition to subsonic values. The inlet on the SR-71 shown in Fig. 13.7 illustrates this design approach.

One might wonder whether the addition of further rocket nozzle hardware to take advantage of the high pressure in the flow could be worthwhile. Indeed, it would, but the mechanical complexity of adding more of a nozzle skirt *while the rocket is in flight* is a challenge that reality might rule out as a solution.

13.6.3 Staging

In practice, this issue is often resolved by staging, i.e., employing a number of series or parallel stages to build a complete rocket. The nozzles can then be designed to operate in an appropriate range of altitudes with acceptable performance. The staging also addresses the concern that a vehicle may be limited in structural strength to a certain level of acceleration. The level of g-forces acting on the structure increases markedly in any ascent to space because the thrust may or may not be reducible as fuel is burned off and the vehicle becomes ever lighter. Further, there is no point to expending power and fuel to haul excess engine thrust capacity and empty tanks to an altitude where they are not needed. To that end, the (parallel) solid propellant boosters (a stage of sorts) of the Space Shuttle were designed to provide a decreasing amount of thrust as they burn during their 125 s of use when they burn out at about 145,000 feet of altitude. The Apollo rocket system is a set of stages in series, meaning they are fired in a sequence after an earlier stage used up its fuel supply.

13.6.4 Specific Impulse and a Little Chemistry

In order to circle back to the discussion of flow along a streamline, a few comments involving the conservation of total enthalpy, are particularly interesting in connection with the function of a rocket engine like the SSME. Expansion of the rocket's jet into vacuum leads, ideally, to a complete conversion of the total enthalpy along the streamlines to kinetic energy. The velocity associated with that kinetic energy is the momentum (per unit mass) wanted for thrust. The more the better, because the only other way to increase the thrust is to increase mass flow rate and that is costly in terms of weight that must be launched. Thus, we examine the only quantities that determine a high jet velocity in chemical rocket: a high total temperature and a low molecular weight of the gas. Rocket propulsion people call this velocity by a special name: thrust per unit *weight* flow rate of propellant, or *specific impulse* (I_{sp}). It is nothing more than the jet velocity divided by the acceleration of gravity. The definition of I_{sp} means that it is measured in seconds.

Hydrocarbon fuels burning with oxygen produce a gas with $I_{sp} = \sim 350$ s. A solid propellant might operate with 280 s. The hydrogen–oxygen combustion in the SSME yields a superior performance with about 450 s ($\sim 14{,}500$ feet/s!) in the vacuum of space but only about 360 s at takeoff where the over-expansion of the flow reduces the performance. Here is a good place to point out that the airbreathing engines that do not have to carry the air (oxygen) used for combustion can be contrasted to a rocket. The specific impulse of a modern turbofan engine in cruise is about 6500 s![3] No wonder we don't use rockets in airliners!

The hydrogen–oxygen reaction is quite energetic yielding high combustion temperatures. The chemically correct proportion of hydrogen and oxygen is $2 \cdot H_2 + 1 \cdot O_2$ to produce two molecules of water vapor. This translates to a ratio of 1:8 by weight. Oxygen is really heavy compared to hydrogen. The resulting jet consisting of pure water vapor has a molecular weight of 18. A maximum total enthalpy per unit mass can be obtained with a reduced temperature and reduced molecular if an excess of hydrogen is processed. Indeed, the SSME burns the reactants *fuel-rich* in 1:6 (H/O) proportion. The molecular weight of the gas is thereby reduced by about 35%. The lower temperature together with the dearth of (chemically very reactive) oxygen ease the design, construction, and reusability of the engine.

We conclude the discussion of rocket flight to space with a comment on dynamic pressure (q) tackled earlier in our story. Recall this pressure involves only speed and air density. Enthusiasts of flights to space might note that launch operations often involve notation of the point along the ascent where the rocket reaches a point of "*max q*". That there is such a maximum should be self-evident as the rocket is flying ever faster into an environment of ever lower atmospheric air density, eventually reaching space with near zero density. The *max q* point is where the aerodynamic heating and aerodynamic loads are at maxima. Later in the flight, these aspects become an ever-smaller concern.

Rocket propulsion involves a lot of chemistry and the identification of "good" fuels and reactions is a part of the challenge. This is part of the work done by rocket scientists! On top of that is the challenge of getting the engine to fire up and run reliably, all without sudden surprises!

One of the interesting dimensions of liquid fuel rocketry is that it employs the same thermodynamic cycle as did the first successful engine: the steam engine. Central to the cycle is compression of a liquid: water in the old and the chemical propellants in the new. Expansion in the old is against a piston or through a turbine. In the new, expansion is via a nozzle to create a propulsive jet. The SSME is a steam engine!

13.7   Airplane Range

Returning all the way back to the discussion of aerodynamic performance of a wing or airplane, we note the aerodynamicist is largely concerned with maximizing the M L/D to obtain long range and low fuel costs in an airliner or similar transport vehicle. That quantity is also central in the goal of a long-range airplane design. If either of

---

[3] Airplane propulsion people use a fuel consumption number called the thrust specific fuel consumption (TSFC) measured in pounds per hour per pound of thrust. TSFC is essentially the inverse of I_{sp}.

the parameters, $M L/D$ or I_{sp}, has an important influence on the empty weight of an airplane, then some serious optimization will have to be undertaken and there lies the heart of successfully designing an airplane.

Epilogue

Friction and the necessary existence of the boundary layer on bodies moving through a fluid make life difficult and challenging for the designer of an airplane. The absence of that reality would make flight impossible.

It is the hope that, in looking over these pages, the subject of flight is better appreciated for the technical achievement that it is. The talent and hard-won insight by many into the complex behavior of air we cannot even see is obviously manifest in the state of flight today. Many theoretical musings and postulates had to be tested by experiments in the laboratory and in flight. The latter often took courage and sometimes ended poorly for the person sitting in the pilot's seat. And yet we, the community of the curious and adventurous, learned from the experiences of those who went before us.

A reality that can never be ignored is that technology can and is always a double-edged sword. It can be used for good and evil. As was the knife that can cut the meat to eat or stab a person, the gun powder that went from fireworks to propel a bullet, so the airplane can bring help to rescue those that need it, help extinguish a fire in a blazing forest, so can it deliver you or me to a distant place. It can also deliver a bomb to an innocent or to a malevolent enemy.

Developing technology is a thrill at the fundamental level but brings with it a need to examine the associated morality. On July 16, 1945, after witnessing the test of the first atomic bomb, Robert Oppenheimer is said to have pondered a piece of Hindu scripture: "Now I am become Death, the destroyer of worlds". Such is the human condition, and the road of life is not easily travelled. We will, however, journey on.

© The Author(s) 2022
R. Decher, *The Vortex and The Jet*,
https://doi.org/10.1007/978-981-16-8028-1

Appendix A
Equations for Quantitative Descriptions

The following is a listing of the equations that may be useful to augment the text discussion. Numerical values of the constant are not provided as these differ between the English and the metric (SI) systems.

A.1 Motion with Circular Streamlines

(Azimuthal) velocity distributions in a vortex and solid body rotation

r is the radial distance from center to streamline,
Vortex: $V_{azimuthal} = \Gamma/2\pi r$; Γ is the vortex strength,
Solid body: $V_{azimuthal} = \omega r$; ω is the angular speed of rotation.

A.2 Air as a Medium

Air is the medium of flight. It consists of about 79% nitrogen, 1% argon and 20% oxygen. The predominance of nitrogen in all air-related processes allows a reasonably accurate description of combustion gas as being like air. While the inclusion of real gas effects is relatively straightforward, it not necessary for the qualitative descriptions of the relevant processes. Further, in an elementary examination of the performance of a gas turbine engine, the fuel mass added to the air stream mass can be neglected.

© The Author(s) 2022
R. Decher, *The Vortex and The Jet*,
https://doi.org/10.1007/978-981-16-8028-1

A.3   The Descriptive Constants

R—ideal gas constant as in the equation of state: $pv = RT$

C_v—specific heat at constant volume, from the internal energy of air: $u = C_v T$

C_p—specific heat at constant pressure, from the enthalpy of air: $h = C_p T$.

[*more precise definitions for the specific heats apply generally, but for our modelling of air, the relations to internal energy and enthalpy as stated are adequate*].

The specific heats C_p and C_v are, in reality, functions of temperature because of the internal complexity of the molecules. However, for modeling purposes and reasonably good descriptions of air in the context of wing and engine performance, taking them to be constant is acceptable. The ideal gas equation of state, above, is very much accurate.

The definition of the enthalpy, h, allows stating: $C_p = C_v + R$

and a definition for the specific heat ratio: $\gamma = C_p/C_v = 1.4$ for air.

A.4   The Gas Properties

Here and in the text, the use of absolute temperatures (T) and pressures (p) is implicit, unless otherwise noted. The carriage of extra reference constants such as atmospheric pressure and a standard temperature is simply too awkward.

The text refers to *static* values of these quantities that are measured by an observer at rest in the medium. Example: a butterfly holding a thermometer and a pressure gauge while carried along by a flow would measure static pressure and temperature.

Total pressure and temperature are the values measured by a stationary observer at a place where a moving fluid is gently brought to rest relative to that observer. The total temperature and total pressure are measured by the butterfly above just before it impacts the leading edge of a wing or similar obstruction that brings the flow to rest.

The total enthalpy, or equivalently the total temperature, is the measure of all the energy in a moving fluid.

The specific volume (cubic meters per kilogram or cubic feet per pound mass, v) and the density (kg/m^3, etc., ρ) are related through $v = 1/\rho$.

A.5   The Flight Equations

Dynamic pressure: $q = \frac{1}{2}\rho V^2$

[the symbol q is conventional and is not related to the same symbol for heat interactions]

Airplane force coefficients: C_L or $C_D = $ Lift or Drag/($\frac{1}{2}\rho V^2$) · (Wing Area).

Propulsive thrust:

[of a system moving at velocity V_0, producing a jet of speed V_j, neglecting the usually small pressure mismatch term]

$$= (Mass\ flow\ rate) \cdot (V_j - V_0)$$

Energy in jet stream:

$$= (Mass\ flow\ rate) \cdot \tfrac{1}{2}(V_j^2 - V_0^2)$$

Propulsive efficiency involving jet (j) and flight (0) speeds:

$$= \frac{2}{\left(1 + \frac{V_j}{V_0}\right)}$$

A.6 The Equations of Aerothermodynamics

A.6.1 The Energy Equation Along a (Steady) Streamline

Definition h_t (total enthalpy) $= h$ (static) $+ \frac{1}{2} V^2$.

From which follows (with the definition of M and speed of sound, see below) Total temperature:

$$T_t = T\left(1 + \frac{\gamma - 1}{2}M^2\right)$$

Energy equation for steady, compressible fluid flow:

$$h_{t2} - h_{t1} = q\text{(heat)} + w\text{(mechanical work)}$$

For adiabatic flow (no heat interaction) and no work interactions:

$$h_t = h + \tfrac{1}{2} V^2 \text{ is conserved}$$

The energy equation for an incompressible fluid (density (ρ) must be constant, Bernoulli principle):

$$p/\rho + \tfrac{1}{2} V^2 \text{ is conserved}$$

A.6.2 Compressibility

Static speed of sound:

$$a = \sqrt{\gamma RT}$$

Mach number:

$$M = \frac{V}{a}$$

In a steady, reversible flow pressure changes lead to density changes:

$$\frac{d\rho}{\rho} = -M^2 \frac{dV}{V}$$

[see Appendix B; $\frac{dx}{x}$ is the fractional change in x].

A.6.3 Isentropic Relation Between Two States 1 and 2

$$\frac{p_2}{p_1} = \left(\frac{T_2}{T_1}\right)^{\frac{\gamma}{\gamma-1}}$$

with $\gamma/(\gamma - 1) = 3.5$ for air.
    From which follows,
    Total pressure:

$$p_t = p\left(1 + \frac{\gamma-1}{2}M^2\right)^{\frac{\gamma}{\gamma-1}}$$

A.6.4 Heating Total Pressure Loss

$$\frac{dp_t}{p_t} = -\frac{\gamma}{2}M^2\frac{dq}{C_pT_t}$$

where dq is the differential amount of heat added.

Appendix B
Some Quantitative Aspects of Aerodynamics and Thermodynamics

B.1 Entropy is a Fact

This appendix, as a discussion of entropy, should properly be the beginning of a study of aerodynamics and thermodynamics, the studies of air and heat in motion. In a text such as this, the result might discourage the reader and the aims of the text would have been missed. It is, however, unavoidable to involve the Second Law of Thermodynamics in an honest discussion of the physics of flight. Further, such a discussion is made much more effective with differential calculus because the second law of thermodynamics is conveniently written in this form.

We have described many fluid motion processes as conserving a quantity called the entropy. The following paragraphs are intended to shed a little more light on that subject using slightly more sophisticated tools.

What is this property called entropy? It is an abstract property whose existence follows directly from the second law of the thermodynamics. The logic parallels the thinking about the first law that is also called an energy conservation principle. The concept of the abstract idea of "energy" as a property originated there. In a similar manner, entropy originates from the second law.

What does the second law of thermodynamics really state? In our everyday experience, we observe that work done, such as dragging a brick across the floor, stirring a liquid, or compressing a gas, is convertible to heat. Thus, heat and work are somehow equivalent. On the other hand, can heat be converted to work? Yes, but within limits. That is at the heart of classical thermodynamics and has to be reconciled with the observation that heat only flows from a hot body to a colder one and not in reverse.

A good book on classical thermodynamics will make the case that heat transferred from whatever we are dealing with (the system) to its external environment (or the reverse) can be done reversibly if the transfer occurs with no fall in temperature between the system and its environment. Without proving it here, this amounts to saying that the differentially small amount of heat transferred divided by some measure of temperature can be made reversible. Briefly, the idea is to show that transferring heat can be made reversible when the temperature of the system being

© The Author(s) 2022
R. Decher, *The Vortex and The Jet*,
https://doi.org/10.1007/978-981-16-8028-1

examined is always at the same temperature as the environment to which it gains or loses heat. This involves arguing for the (unrealistic) existence of a large (infinite?) array of 'reservoirs' each at slightly different and fixed temperatures. While this is not practical, it is possible in a thought experiment.

The simplest way to express that mathematically is to make the measure of temperature, the temperature itself. Thus, the process dq/T can be made reversible.

The general form of the First Law is

$$\partial q = du + \partial w$$

the symbols q and w describe heat and work interaction with a system, which for our purposes, could be a volume of air. In a classical statement, the differential "∂"s associated with heat and work are given a modified symbols (to contrast with "d") to indicate that a *process* is described and, specifically, that q and w are not properties. On the other hand, u, the internal energy (generally dependent on temperature and sometimes even pressure), is a property.

When the first law is written in a reversible form, work done must be written with $\partial w = pdv$. This statement is the mechanical equivalent of work expressed as force times distance.

Thus, one can write

$$\left(\frac{dq}{T}\right)_{rev} = \frac{du}{T} + \frac{p\,dv}{T}$$

The right-hand side of this equation involves only properties, hence the left-hand side must also be a property. It is, in fact, a differential of the entropy, ds.

For an ideal ($pv = RT$) gas, this reads

$$\left(\frac{dq}{T}\right)_{rev} = ds = \frac{du}{T} + R\frac{dv}{v} = C_v\frac{dT}{T} + R\frac{dv}{v}$$

In differential form, the ideal gas law can be written as

$$\frac{dp}{p} + \frac{dv}{v} = \frac{dT}{T}$$

so that the entropy can finally be written in terms of the parameters that really interest us, namely pressure and temperature. Before we do, we will focus on air which is also a perfect gas with constant specific heats. For that case, the definition of enthalpy gives

$$C_p = C_v + R$$

and the differential entropy becomes

$$\frac{ds}{C_v} = \frac{dT}{T} - \frac{\gamma - 1}{\gamma}\frac{dp}{p}$$

using the definition of γ. This is the origin of the isentropic ($ds = 0$) relation between pressure and temperature when integrated for a process between two states.

B.2 Compressibility

While we are immersed in the differential forms of the equations governing various processes, it is also convenient and fairly straightforward to examine the criterion expressed in Chap. 5 to determine whether a flow process involves a gas that must be described as compressible or whether the incompressible (Bernoulli) formulation is acceptable.

To do this we invoke Newton's second law of motion (simply stated as $F = ma$) expressed here in differential form applied to an element of air. This relation (often called a momentum equation) reads, for a unit volume and in the absence of friction forces, as

$$dp + \rho\, V\, dV = 0$$

For an incompressible fluid, this relation leads directly to the Bernoulli principle. More generally, this relation can be rewritten in terms of the very convenient fractional changes as

$$\frac{dp}{p} = -\frac{\rho\, V^2}{p}\frac{dV}{V} = -\gamma M^2\frac{dV}{V}$$

With the speed of sound stated as

$$a^2 = \frac{\gamma p}{\rho} = \gamma\, RT$$

Thus, the Mach number appears as the important parameter to state whether velocity changes are sufficient to change the pressure of the air in a significant way. With

$$\frac{p_2}{p_1} = \left(\frac{p_2}{p_1}\right)^{-\gamma} \text{ or } \frac{dp}{p} = -\gamma\frac{d\rho}{\rho}$$

valid for isentropic processes, the density changes are easily calculated from the velocity changes:

$$\frac{d\rho}{\rho} = -M^2 \frac{dV}{V}$$

showing the important role that Mach number plays in determining whether a flow may be modeled as incompressible when it might not be.

B.3 Boundary Layers

The momentum equation as a form of Newton's second law of motion is a vector equation with associated directions. If that equation is applied to the direction normal to a surface along which there is flow, we note that there can be little or no flow acceleration in that direction because the boundary does not allow it. It is therefore a safe conclusion to state that the static pressure normal to the surface, i.e., through the boundary layer is uniform in that direction. That is the observation that allows the interpretation that measurements made on the surfaces are accurate reflection of the pressure in the inviscid flow field outside the boundary layer.

Appendix C
Induced Drag

An overview of the airplane designer's concern with lift induced drag (Fig. C.1).

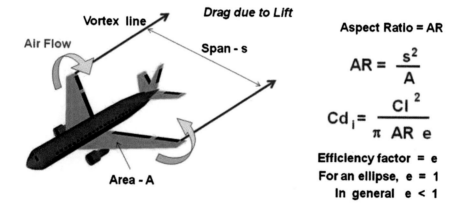

Pressure difference from top to bottom of the wing causes spillage around the wing tips.

Downwash from the tips induces local angle of attack with additional drag component on a finite wing.

Fig. C.1 This NASA chart illustrates well the role of wing aspect ratio (AR) and its role in determining drag due to lift. A is the wing area, s the wingspan, and C_L is the wing lift coefficient. (NASA)

© The Author(s) 2022
R. Decher, *The Vortex and The Jet*,
https://doi.org/10.1007/978-981-16-8028-1

Appendix D
Aerodynamic Performance Summary

The following graphic is a so-called *High Speed Drag Polar* for a Boeing 727–100. The figure is like Fig. 7.12 except that the axes are switched in orientation. The origin is not shown because only the information of practical interest is displayed. Note the significant drag increase as flight Mach number is increased. On several curves, the points of the maximum *L/D* is identified (s. D.1).

From the plot, the values for maximum *L/D* are roughly as indicated in Table D.1. Thus, for maximum aerodynamic efficiency, the best flight Mach appears to be near 0.70. The table also notes the values of *M L/D* for maximum range. That combination of measure maximizes nearer 0.80. Fortunately, the maxima in these two estimated parameters vary slowly with Mach number so that there is some operational flexibility.

Table D.1 Variation of lift to drag ratio and *M L/D* with Mach number for the Boeing 727–100

Mach No.	max *L/D*	*M L/D*
0.7	16.5	11.5
0.8	15.1	12
0.85	12.8	10.9
0.88	10.3	9

© The Author(s) 2022
R. Decher, *The Vortex and The Jet*,
https://doi.org/10.1007/978-981-16-8028-1

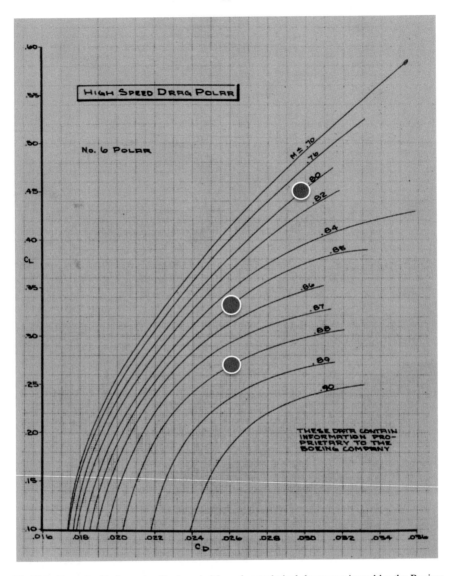

Fig. D.1 From the Performance Engineer's Manual, a technical document issued by the Boeing Company about the Model 727–100 in 1967, a plot of the airplane drag polar: Airplane lift versus drag coefficient for various flight Mach numbers. The circles are approximate locations of maximum *L/D* (The Boeing Company)

Glossary of Technical Terms

Adiabatic Description of a process that involves no heat transfer between a fluid and its environment.

Adverse pressure Flow with pressure increasing in the flow direction.

Airfoil Geometric description of a surface to deflect air flow.

Bernoulli Mathematical relation between pressure and flow speed for an incompressible fluid.

Boundary layer Fluid layer flowing adjacent to a boundary.

Bypass Compressed propulsion air not used in the engine itself.

Camber Descriptive quantity measuring an airfoil's shape.

Choking Flow limitation by an internal flow area where local Mach number is 1.0.

Chord Length of an airfoil in the flow direction.

Circulation Sum of velocities around a closed loop.

Compressibility Susceptibility of a fluid's density to changes in pressure.

Compression Process of increasing the pressure of a fluid or, equivalently for a gas, decreasing its density.

Compressor Device using mechanical power to compress air.

Condensation Vapor phase change to liquid state.

Constant pressure combustion Combustion carried out in a steady flow process.

Constant volume combustion Combustion carried out as an explosive heat release in a piston/cylinder (engine) during the time piston is near top dead center.

Contrails Condensed combustion water vapor trail in the atmosphere after passage of an airplane.

Downwash The mass of air pushed downward by a wing or its downward velocity.

Drag polar A plot of lift versus drag for an airfoil or an airplane.

Dynamics The branch of mechanics concerned with the motion of bodies under the action of forces (see also "kinematics").

Enthalpy Measure of thermal energy in a flowing medium, a property.

Entropy Material property of a substance whose changes during a process reflect the degree of reversibility of those changes.

Expansion Process of allowing a gas to increase in volume.

Fan Compression device for air not used in the engine.

© The Author(s) 2022
R. Decher, *The Vortex and The Jet*,
https://doi.org/10.1007/978-981-16-8028-1

Flow separation A failure of flow to follow the intended contour of a surface.

Friction The irreversible process of work converted to heat.

Indicated air speed Speed inferred from a measurement of total and static pressures by means of a pitot probe.

Induced drag Drag due to lift.

Internal combustion engine An engine operating with pistons within cylinders to produce power.

Internal energy The energy associated with the motion of atoms or molecules and their molecular configuration.

Irrotational flow A flow that has not been subject to rotational forces (such as friction).

Jet Fluid that moves faster relative to its environment in a specific direction.

Kinematics The branch of mechanics concerned with the motion of objects without reference to the forces which cause the motion (see also "dynamics").

Kinetic energy Energy associated with motion.

Leading edge Front edge of a wing or airfoil.

Mach number Ratio measuring air speed relative to its wave propagation speed.

Momentum A physical measure of the speed and direction of a mass.

No-slip condition The reality that fluid molecules in contact with boundary surface molecules are in dynamic equilibrium.

Pitot probe A device for measuring total and static pressure.

Ram pressure Another word for total pressure.

Reynolds number A non-dimensional parameter useful for determining the importance of viscous shearing forces relative to dynamic forces.

Shock wave A stationary wave formed under circumstances where information transmission upstream by sound waves is not possible because the flow velocity is larger than wave propagation speed.

Specific impulse Thrust per unit weight flow rate.

Stagnation Process of bringing flow to rest.

Static pressure Pressure measured by an observer at rest in a fluid medium.

Thermal energy Energy associated with random kinetic energy of fluid atoms and/or molecules.

Total pressure Pressure realized by bring flow to rest (also stagnation pressure or ram pressure).

Total temperature Temperature realized bring flow to rest (also stagnation temperature).

Trailing edge Rear edge of an airfoil or wing.

Turbine Vaned rotating wheel designed to extract mechanical power from a gas stream.

Vortex A macroscopic fluid element with a velocity distribution in a circular direction that varies inversely with distance from the center.

Vortex sheet A planar distribution of vortices.

Vorticity A spatial distribution of vortices.

Wake The boundary layer air behind a body moving through a fluid.

References

Alfredsson, P.H., and M. Matsubara. Free-stream turbulence, streaky structures and transition in boundary layer flows. AIAA Paper 2000–2534.
Van Dyke, M. 1983. *An Album of Fluid Motion.* Parabolic Press.

Bibliography

Abbott, I.A., and A.E. von Doenhoff. 1949, 1959. *Theory of Wing Sections.* Dover.
Abbott, I.A., A.E. von Doenhoff, and L.S. Stivers. 1945. *Summary of Airfoil Data.* NACA TR-824.
Alfredsson, P.H., and M. Matsubara. Free-stream turbulence, streaky structures and transition in boundary layer flows. AIAA Paper 2000–2534.
Anderson, J. 2016. *Introduction to Flight*, 8th ed. New York: McGraw-Hill Education.
Decher, R. 1994. *Energy Conversion.* Oxford University Press.
Decher, R. 2020. *Powering the World's Airliners.* UK: Pen & Sword.
Decher, R. 2022. *Turbulent Journey - The Jumo Engine, Operation Paperclip, and the American Dream.* Schiffer Publishing.
Glauert, H. 1947. *The Elements of Aerofoil and Airscrew Theory.* Cambridge University Press (originally 1926).
Golley, J. 1996. *Jet.* Datum Publishing Limited.
Kerrebrock, J.L. 1992. *Aircraft Engines and Gas Turbines.* Cambridge, MA: MIT Press.
Liepmann, H.W., and A. Roshko. 1957, 1985, 1993. Elements of gas dynamics. *American Institute of Physics.*
Oates, G.C. 1984. *Aerothermodynamics of Gas Turbine and Rocket Propulsion.* New York: American Institute of Aeronautics and Astronautics.

© The Author(s) 2022
R. Decher, *The Vortex and The Jet*,
https://doi.org/10.1007/978-981-16-8028-1

Index

Printed in the United States
by Baker & Taylor Publisher Services